Entrepreneurship Opportunities in Fisheries and Aquaculture

NIPA® GENX ELECTRONIC RESOURCES & SOLUTIONS P. LTD.
New Delhi-110 034

About the Editors

Dr. B. P. Mohanty, Asst. DG (Inland Fisheries) ICAR since March 2020. MVSc (Biochemistry) from IVRI, Izatnagar and PhD from JNU, New Delhi. Started his career as Asst. Professor (Biochemistry) in OUAT, Bhubaneswar in Mar 1990. Joined ARS, ICAR in 1994. Served as Scientist, Senior Scientist, Principal Scientist and Head of Division at CIFRI, Barrackpore. Conferred Biotechnology Overseas Associateship Award (BTOA) by DBT, Government of India for advanced training in Proteomics. Published more than 100 research papers in journals of repute and has three Books, 10 Research Bulletins, two Open-access Database (NutriFishIN & FISHPROT) and one Mobile App (NutrifishApp) to his credit. Life Member and Fellow of prestigious scientific societies and academies. Fellow of National Academy of Agricultural Sciences (FNAAS), Zoological Society Kolkata (FZS), Inland Fisheries Society of India (FIFSI), Zoological Society of India (FZSI); Hon. Fellow of International Society for Ecological Communications (FISEC) and Life Member NASI Allahabad. Received Gold Medal from NAARM, Hyderabad; Zoological Society of India and OP Odum Gold Medal from FISEC & Lifetime Achievement Award in Biochemistry. Member Editorial Advisory Board, CAB International (CABI), UK. Member Board of Management, OUAT Bhubaneswar, Odisha & TNJFU Nagapattinum, Tamil Nadu. Dr. Mohanty is currently the President of Society of Veterinary Biochemists and Biotechnologists of India (SVBBI).

Dr. Shahaji Phand, Deputy Director and Centre Head of Centre for Extension in Agri-allied Sector and also Principal Coordinator, Agri-Clinics and Agribusiness Scheme (AC&ABC), National Institute of Agricultural Extension Management (MANAGE), Hyderabad. He is veterinary graduate and completed Bachelor of Veterinary Science and Animal Husbandry (B. V. Sc. A.H) degree in 2002 from Maharashtra Animal and Fishery Science University, Nagpur Maharashtra. He completed his Master and Doctoral Degree in Veterinary Extension Education from Indian Veterinary Research Institute, Izatnagar, Bareilly, Uttar Pradesh in 2005 and 2008 with Junior and Senior Research Fellowship respectively. After completion of doctoral degree, he joined as Assistant Professor in Nanaji Deshmukh Veterinary Science University (NDVSU) in June 2008, promoted as Associate Professor in 2012 in same university. After serving 8 years as faculty in university, he joined as Assistant Director in MANAGE, Hyderabad in 2016 and promoted to post of Deputy Director in 2018. He has developed 02 technologies in the form of computer software's namely Animal

Health Information System (AHIS) and Health Information System for Dairy Animal (HIS) as part his doctoral research, which is patented and commercialized. For this contribution he has been conferred with Dr. C.M. Singh Award for Best Ph. D Research at national level. He has published more than 30 research and review papers in national and international journal and edited of more than 27 books. He has been editor, reviewer and member of editorial board of national and international Journals. He has organized and coordinated more than 300 training programs for senior and middle level officers of State Department of Agriculture and Allied sector, Scientist, SAU faculties, Students etc. which benefit reached more than 4000 stakeholders. He has organized and participated national level conferences, seminar, workshop and contributed for policy formulation. As a Principal Coordinator, he is implementing Agri-Clinics and Agribusiness Scheme (AC&ABC) with the help of 126 Nodal Training Institute (NTIs) across the country, which have trained more than 7500 agricultural graduates and diploma holder, among which more than 3000 trained candidates have established their agri-venture in last two years and serving farming community.

Dr. Sushrirekha Das presently working as MANAGE Fellow in extension in Agri-allied Centre, MANAGE, Hyderabad, India in the Fishery sector. She has completed her Graduation in Zoology as a topper from Swami Vivekanda Autonomous College, Jagatsinghpur, completed her M.Sc in Zoology with Specialization in Fisheries Science and did her Ph.D. in Zoology with research specialization in Fisheries Science. Her core research is in probiotics from Sriram Chandrabhanj Deo University, Mayurbhanj, Baripada, Odisha. She has put 3 Years of service in teaching as Zoology lecturer. She is having 5 years as Fishery extension officer. She is having more than 8 years of experience in the field of education, research and extension in aquaculture sector. She has submitted 15 numbers of genes to the NCBI Gen Bank. She has published patent in Probiotic rich fermented Amla (Embilica officinalis) fruits and a method of preparation thereof. She has developed 4 probiotics aquaculture products. She has guided 4 nos. of Master degree and graduate students, and handled one Research Projects in SERB, India. She has organized and coordinated more than 70 Workshops, Training Programmes on aquaculture and animal husbandry sector by MANAGE. She has trained more than 500 fish and Shrimp farmers of Odisha and other states through various funding agencies such as NFDB and MANAGE. She has coordinated 115 nos. of SCSP training program in the Fy- 2021-22, there are 4805 fishermen, fisher women, aquapreneurs, students and research scholars in PAN India. She has published more than 20 Research papers of National and International Repute and more than 8 popular articles. She has published 22 Chapters in different books.

She has contributed as editor in 26 e-books in agricultural extension. She has received many prestigious Awards such as Young Scientist Award, Best poster/oral presentation Award, Best Performer Award and she has awarded with best service provider award in fishery extension service and Rastriya Samaj Seva Ratna Award etc.

Entrepreneurship Opportunities in Fisheries and Aquaculture

B.P. Mohanty
Shahaji Phand
Sushrirekha Das

nipa®

NIPA® GENX ELECTRONIC RESOURCES & SOLUTIONS P. LTD.
New Delhi-110 034

NIPA® GENX ELECTRONIC RESOURCES & SOLUTIONS P. LTD.

101,103, Vikas Surya Plaza, CU Block
L.S.C.Market, Pitam Pura, New Delhi-110 034
Ph : +91 11 4386 0225, 9717133558, 9540816132
E-mail: newindiapublishingagency@gmail.com
Website: www. nipresources.com

Print ISBN: 978-93-58878-76-9

ebook ISBN: 978-93-58876-45-1

Composed and Designed by NIPA®.

Dedicated to
Our Families

सत्यमेव जयते

डॉ. हिमांशु पाठक
Dr. Himanshu Pathak
सचिव (डेयर) एवं महानिदेशक (आईसीएआर)
Secretary (DARE) &
Director General (ICAR)

भारत सरकार
कृषि अनुसंधान और शिक्षा विभाग एवं
भारतीय कृषि अनुसंधान परिषद
कृषि एवं किसान कल्याण मंत्रालय, कृषि भवन, नई दिल्ली–110 001

GOVERNMENT OF INDIA
DEPARTMENT OF AGRICULTURAL RESEARCH AND EDUCATION (DARE)
AND
INDIAN COUNCIL OF AGRICULTURAL RESEARCH (ICAR)
MINISTRY OF AGRICULTURE AND FARMERS WELFARE
Krishi Bhavan, New Delhi 110 001
Tel: 23382629 / 23386711 Fax: 91-11-23384773
E-mail: dg.icar@nic.in

Foreword

The blue economy, which encompasses sustainable use of ocean resources for economic growth, presents significant opportunities for India. Fisheries and aquaculture can contribute for economic growth by providing employment, income generation, and economic development, particularly in coastal areas. By intelligently harnessing the potential of fisheries and aquaculture, India can not only enhance food and nutrition security but also contribute to broader socio- economic development goals while ensuring the long-term sustainability of its aquatic resources. The aquatic agri-food systems are playing a big role, and can still play a much bigger role, in achieving global nutrition security and the sustainable development goals (SDGs).

Hunger and malnutrition are serious challenges and ‘Blue food’ and ‘Blue economy’ would have tremendous contribution towards the control of hunger in all forms; protein hunger, calorie hunger and hidden hunger. For this, there is need to upscale the activities at a much bigger level to give a big momentum. It is now realized that entrepreneurship approach is needed for taking fisheries and aquaculture to ‘industry scale’. Entrepreneurship is the ability and readiness to develop, organize and run a business enterprise, with all its uncertainties, to make it a profitable venture and the most prominent example of entrepreneurship is the starting of new businesses. Keeping in view, the much-desired need for entrepreneurship in fisheries and aquaculture, this book “Entrepreneurship Opportunities in Fisheries and Aquaculture” has been designed. The different sub-sectors including hatchery for fish breeding and culture, cage farming, pen culture, fish feed, ornamental fish and aquarium industry, dry-fish and fish processing sector, pearl culture, mariculture, aquaponics, fishing crafts and gears, sports fisheries and aqua-tourism, where there is a scope for entrepreneurial venture have been discussed to attract the bright minds with passion for entrepreneurship in aquaculture.

I appreciate the efforts of the authors and hope the document would serve as a ready reckoner for entrepreneurship opportunities in fishery sector.

Dated the 19th March, 2024
New Delhi

(Himanshu Pathak)

डॉ. पी. चन्द्रा शेखरा
महानिदेशक

Dr. P. Chandra Shekara
Director General

Message

The fisheries sector has long been a cornerstone of global economies, providing sustenance, employment, and economic prosperity to countless communities. However, in today's dynamic world, the sector is experiencing a wave of transformation driven by technological advancements, changing consumer preferences, and environmental concerns. In this changing scenario the stakeholders are searching entrepreneurship opportunities in fisheries and aquaculture sector.

This book serves as a compass, guiding aspiring entrepreneurs, seasoned fishermen, and industry stakeholders alike toward lucrative ventures and sustainable practices. From innovative aquaculture practices to value-added seafood products, from eco-friendly fishing gear to digital marketing strategies. The book entitled "Entrepreneurship Opportunities in Fisheries and Aquaculture" offers a comprehensive roadmap for entrepreneurs to navigate the complexities of the industry and harness its vast potential.

This book delves into the multifaceted dimensions of fisheries, exploring the intricacies of their biology, ecology, management, and socio-economic implications.

I commend the authors for their dedication and expertise in bringing this vital subject to light, and I extend my gratitude to all editors and those who contributed to this endeavour. Together, let us embark on a journey towards a future where our fisheries thrive in harmony with nature, benefiting both present and future generations.

Dr. P. Chandra Shekara
Director General, MANAGE

राष्ट्रीय कृषि विस्तार प्रबंध संस्थान (मैनेज)
(कृषि एवं किसान कल्याण मंत्रालय, भारत सरकार का एक स्वायत्त संगठन, राजेन्द्रनगर, हैदराबाद - 500 030. तेलंगाना, भारत.)
NATIONAL INSTITUTE OF AGRICULTURAL EXTENSION MANAGEMENT (MANAGE)
(An Autonomous Organization of Ministry of Agriculture and Farmers Welfare, Government of India)
Rajendranagar, Hyderabad - 500 030. Telangana State. INDIA
Telephone : +91 040-2459 4505, 24015253, Fax: +91-(040) 24015388
E-mail : dgmanage@manage.gov.in, Web : www.manage.gov.in

Preface

Fish is a super-food. It is one of the best sources of quality animal proteins besides the polyunsaturated fatty acids (PUFAs), especially the omega-3 PUFAs EPA and DHA, and micronutrients. Blue revolution has made aquaculture one of the fastest growing food production sectors. The, 'Blue food' and 'Blue economy' holds a great promise for the future by providing food and nutritional security to the ever-growing world population.

India being a highly populous country, there is a need for scaling-up of the fish production for ensuring food and nutrition security and livelihood. After formation of the Ministry of Fisheries, Animal Husbandry and Dairying (MoFAD), the Department of Fisheries (DoF), Government of India has taken several new initiatives for boosting up the fisheries and aquaculture sector. The Pradhan Mantri Matsya Sampada Yojana (PMMSY) is a big step in that direction. It is now realized that entrepreneurial approach is needed for taking fisheries and aquaculture to 'industry scale'. Entrepreneurship is the ability and readiness to develop, organize and run a business enterprise, with all its uncertainties, to make it a profitable venture and the most prominent example of entrepreneurship is the starting of new businesses. It could be of any type; small business entrepreneurship, scalable start-up entrepreneurship, large company entrepreneurship, and social entrepreneurship.

Keeping in view, the much desired need for entrepreneurship in fisheries and aquaculture, this book "Entrepreneurship Opportunities in Fisheries and Aquaculture" has been designed. The different sub-sectors including hatchery for fish breeding and culture, cage farming, pen culture, fish feed, ornamental fish and aquarium industry, dry-fish and fish processing sector, pearl culture, mariculture, aquaponics, fishing crafts and gears, sports fisheries and aquatourism, where there is a scope for entrepreneurial venture have been discussed to attract the bright minds with passion for entrepreneurship in aquaculture.

There is a focus on entrepreneurship in fisheries and aquaculture. The book provides ideas on avenues in aquaculture for starting business enterprise. This is expected to be an ideal handbook for planners, aquaculture practitioners, funding agencies and Dreamers who are passionate about Industry-scale aquaculture.

The Editors

Contents

1

Entrepreneurship Opportunities in Fisheries and Aquaculture Sector in India

B.P. Mohanty and Yasmeen Basade

Fisheries Science Division - Indian Council of Agricultural Research
Krishi Anusandhan Bhawan II, Pusa, New Delhi-110012

Abstract

This chapter provides an overview of fisheries sector in India and discusses the scope and opportunities for entrepreneurship in different sub-sectors of fisheries and aquaculture in the country.

Keywords: *Aquaculture, Fisheries, Entrepreneurship opportunities, Ornamental fisheries, Marine fisheries, Coldwater fisheries*

1. Introduction

India completed its 75 years of independence in the year 2022. Government of India celebrated the country's journey across its 75 years of independence and its magnificent achievements during that period as 'Azadi Ka Amrit Mahotsav'. The official journey of Azadi ka Amrit Mahotsav commenced as a 75-week long countdown from 12th March 2021 up to the 75th anniversary of India's independence on 15th August 2022 and continued for a year beyond up to 1^{5th} August 2023. The Azadi Ka Amrit Mahotsav was commemorated as a Jan-Utsav in the spirit of Jan-Bhagidari, which means, a public festival with public participation.

The Hon'ble Government of India has, through this campaign envisioned the foundation of a self-reliant India and he said, "*The Azadi Ka Amrit Mahotsav means elixir of energy of independence; elixir of inspirations of the warriors of freedom struggle; elixir of new ideas and pledges; and elixir of Aatmanirbharta. Therefore, this Mahotsav is a festival of awakening of the nation; festival of fulfilling the dream of good governance; and the festival of global peace and development*"[1].

Azadi ka Amrit Mahotsav paved way to envisage a vision for the country, set goals and realize where the country would be in 2047, when India marks 100 years of its independence. Government of India termed the journey from 75 years to 100 years, the next 25 years, as the "*Amrit Kaal,*" which means a most auspicious period to start a new venture, signifying that it's a time to make resolutions and achieve them (*Sankalp se Siddhi*)[2].

In the second quarter of the financial year 2023-24, the success of 18th G20 Summit, Chandrayaan-3 and Mission Aditya have instilled confidence and vigour through the entire nation across every sector including trade and industry. Government of India has set goals to make India a fully developed nation by 2047, achieve a $5 trillion economy by 2026, and become the world's third-largest economy by 2027. To achieve these three goals, the industry will need to play a significant role[2]. Here it is pertinent to acknowledge that in India agriculture plays a pivotal role and forms mainstay of the country's economy. The share of Gross value added (GVA) of agriculture and allied sectors in total economy at current prices was 18.3 per cent during 2022-23[3]. Also, it provides employment to around 45 per cent of the total workforce[4].

Agriculture is one of the major economic activity which adds to the overall wealth of the country. Bulk of the agriculture economy comes from the contributions of small and marginal farmers; however, these. But these farmers face the challenges of unfavourable weather conditions, lack of adequate quality inputs, absence of timely advisory, lack of access to better markets, lack of investments and many other. In the current scenario of government's economic development policies, entrepreneurship development in the field of agriculture and its allied activities is one of the best alternatives to find employment opportunities, generate income, assure food and nutritional security and contribute to the national economy.

2. Fisheries and Aquaculture Sector

India has vast fishery resources extending from the Himalayas up to the Indian Ocean. The marine fisheries resources are spread along the vast coastline of 8118 km, 2.02 million square km of Exclusive Economic Zone (EEZ), and 0.53 million square km of continental shelf area. To harness these resources, along the coastline there are 3461 fishing villages and 1457 landing centres. On the other hand, the inland fisheries resources are in the form of freshwater bodies which are the rivers and canals- 2.80 lakh km, small reservoirs- 7.54 lakh hectares, medium and large reservoirs- 20.45 lakh hectare, ponds and tanks- 24.48 lakh hectares, beels/ oxbow lakes- 13.55 lakh hectares and brackishwater bodies- 10.66 lakh hectares[5].

These fishery resources harbour varied species of finfishes, aquatic invertebrates such as shellfishes/ crustaceans, molluscs, coelenterates, etc., and aquatic weeds which have potential value for fishing, aquaculture and product development. The total aquatic genetic resources of India comprise 9456 species. Of these 2508 native and 291 exotic finfish species have been catalogued. Among native finfishes 1518 are marine, 877 are freshwater and 113 are brackishwater species. Besides finfishes, the aquatic germplasm resources include 2,934 crustacean species (2,430 marine and 504 freshwater species), 5,070 molluscs species (3,370 marine and 1,700 freshwater species), 765 echinoderm species, 486 sponge species and 844 seaweed species[6].

In India fisheries continued to be one of the fastest growing sectors. Fish production increased from 0.75 million tonnes in 1950-51 to 16.248 million tonnes in 2021-22, of this total fish production 12.121 million tonnes came from inland region and 4.127 million tonnes from marine zone. India is the third largest fish and aquaculture producing country in the world, accounting 8% of global fish production. Correspondingly, India is also the third largest fish exporting country. In 2021-22 exported 1.36 million tonnes of seafood worth USD 7.76 billion. Fisheries contributed about 1.1% to Indian economy and about 6.72% to agriculture economy. The sector provides livelihood to more than 28 million fishers and fish farmers at the primary level and many more along the value chain including a large number of economically disadvantaged population and vulnerable communities[5]. With the available vast and varied fisheries resources in the form of marine and inland water bodies, fisheries is a promising sector with immense growth potential. This sector offers several entrepreneurship opportunities in the field of capture and culture fisheries and its related subsidiary activities. Some of the fisheries technologies which can be considered as entrepreneurship options are described hereunder.

3. Entrepreneurship Development Opportunities in Fisheries

In fisheries a large number of opportunities exist to start a business venture, both in the inland sector as well as marine sector and in both the fields of capture and culture fisheries activities. Such as fish farming, seed production, feed production, fish processing, fish by-products production, freezing, drying, curing, cold chain, packaging, neutraceuticals, transportation, marketing, craft and gear, navigation, and other auxiliary activities as discussed below:

3.1 Culture fisheries

Fish farming

Fish farming or fish culture is a very vast field, it mainly involves growing of fish/ shellfish in enclosures to be sold as food. The enclosures can be either ponds, raceways, reservoirs, wetlands, cages or pens. The fishes cultured can be single species, monoculture or multispecies, polyculture. The fish culture system could be a stand-alone system or integrated with either crops like paddy, horticulture plantation or livestock such as poultry, duckery, cattle and pig farming. Presently, in our country about seventy-five per cent of the fish produced comes from these inland culture systems. The inland fish farming is undertaken in the freshwater and brackishwater systems. Marine fish farming and/or mariculture in undertaken in open sea cages as well as in the estuarine, brackish, coastal, and offshore waters.

Fig. 1.1: Integrated farming System (IFS) at ICAR- IARI, New Delhi

ICAR- IARI, New Delhi

In freshwaters finfish, mainly carp, culture is dominant. The major contributors of finfish aquaculture production are the three Indian major carps (catla, *Catla catla*; rohu, *Labeo rohita* and mrigal, *Cirrhinus mrigala*) 6.297 million tonnes and three exotic carps (silver carp, *Hypophthalmichthys molitrix*; grass carp, *Ctenopharyngodon Idella* and common carp, *Cyprinus carpio*) 1.433 million tonnes. This is followed by catfishes (including the exotic catfish, *Pangasianodon hypophthalmus*) 0.551 million tonnes and minor carps 0.532 million tonnes[5]. In recent past, freshwater aquaculture has been diversified with addition of many more medium and small carps, catfishes and other regionally important species. Some of the species from North-East and Peninsular part of India of aquaculture significance are *Labeo dussumieri, L. kontius, L. dero, L. dyochilus, Tor khudree, T. mussullah, Neolissochilus hexagonolepis, N. wynaadensis, Semiplotus semiplotus, Osteobrama belangeri, Clarias dussumieri, Gonoproktopterus* (*Hypselobarbus*) *curmuca, C. cirrhosa, Puntius pulchellus, P. sarana, P. carnaticus, Channa diplogramme,* and *E. suratensis*. Further, captive breeding and larval rearing techniques have also been developed for *Labeo dussumieri, L. gonius, Gonoproktopterus* (*Hypselobarbus*) *curmuca, E. suratensis, Osteobrama belangeri, Clarias dussumieri, Puntius pulchellus* and *P. carnaticus*[7]. These indigenous species having consumer acceptability and culture potential are being promoted for diversification of aquaculture. Freshwater prawn, *Macrobrachium rosenbergii* is gaining popularity under mono- and poly-culture systems.

Fig. 1.2: Pen culture in wetlands

In recent years ornamental fish farming and pearl culture are also being considered as a promoted as highly promising entrepreneurship initiatives with less initial investments.

Fig. 1.3: Omamental fish farming

In brackishwater culture system shrimps, oysters, mussels, crabs, lobsters, sea bass, groupers, mullets, milk fish, cobia, silver pompano, pearl spot, ornamental fishes and weeds are being cultivated. Of these, promising cultivable species are the shellfish species- *Penaeus indicus, P. monodon, P. merguiensis, P. pencillatus, P. japonicus, P. semisulcatus* and the exotic shrimp *Litopenaeus vannamei*; finfish species- seabass (*L. calcarifer*), grey mullet (*M. cephalus*), pearl-spot (*E. suratensis*), milkfish (*C. chanos*) and mud crabs (*Scylla serrata* and *S. tranquebarica*)[8].

Fig. 1.4: Cage culture in brackishwater areas

Fig. 1.5: Cage culture in reservoirs

In marine waters finfish, shellfish and seaweeds are cultivated of which the commercially cultured species are of shrimps, pearl oysters, mussels, sea cucumber, crabs, lobsters, seabass, groupers, snappers, cobia, mullets, milk fish, rabbit fish, ornamental fishes, seaweeds, etc.

Fig. 1.6: Sea cage farming

Fish seed production

Quality fish seeds are one of the chief components of a profitable fish farming system. At present, there are 2198 eco-carp hatcheries and 417 fiberglass reinforced plastics (FRP) carp hatcheries operating in India[9]. With more and more farmers opting for fish farming there is a growing demand for fish seed. During 2020–21 the total fish seed production in India was 54069 million fries which is insufficient to meet the projected target production of 22 million tons by 2025[5 and 9]. Shrimp seed production is dominated by *P. monodon* and *L. vannamei* species. Presently the Coastal Aquaculture Authority (CAA) registered hatcheries for import of Specific Pathogen Free (SPF) *L. vannamei* broodstock for seed production are 314 and nauplii rearing hatcheries (NRH) are 148 in numbers[10 and 11]. Hence, fish seed production can be a rewarding entrepreneurship option. For fish/ shellfish seed production, brooder fishes and hatchery unit are the major requisites in addition to land and water. Fish seed production can be initiated in small land area with less capital investments and easy returns can be obtained in less duration of time. Fish seed production can make the farmer self-reliant and create employment opportunity for others.

Fig. 1.7: FRP finfish/Shellfish hatchery

Infrastructure development and management for fish farming associated activities

Fish farming encompassed various activities which also includes infrastructure development such as fish farm construction along with water, power, store and amenities; construction and installation of cages and pens for fish and fish seed rearing, hatcheries for seed production, feed mill to manufacture fish feed from locally available ingredients; setting up testing laboratories for soil and water quality analysis, fish diseases diagnostics, etc.; providing crafts like boats to move around in the water bodies and gears like fishing nets to construct cages and pens besides using them to harvest fish, tanks, aquarium and hapa for holding fish and establishing ice plant/ cold storage, fish transport and market facilities and various other related facilities. These infrastructure development, management and maintenance of fish farming associated activities also require a host of professionals who can take up these activities as entrepreneurial ventures.

3.2 Capture fisheries

Marine fisheries form the major portion of the capture fisheries in our country. Export earnings from marine fisheries are growing and hence, is an important contributor to the country's economy. In the financial year 2022-23 India exported 1.735286 million tonnes of seafood worth Rs. 63969.14 crore equivalent to USD 8.09431 billion, increasing 26.73 per cent in terms of quantity and 4.31 per cent in terms of value over year 2021-22[12]. Out of the total fisherfolk population of 28 million in India 4.95 million are engaged in marine fisheries. There are 1.22 million full time, 4.92 lakh part time, 78323 occasional, 2.08 lakh unspecified and 64135 engaged in deep sea fishing[5].

In India marine sector has multi-species fishery which are exploited by various types of crafts and gears. The estimated marine fish landings along the coast of the mainland of India during 2022 was 3.49 million tonnes, showing an increase of 14.53% compared to the landings in 2021. Of the total landings mechanized fishing vessels accounted for 2.85 million tonnes (82.0%), motorized fishing crafts 0.61 million tonnes (17.0%), and non-motorized fishing crafts 0.04 million tonnes (1.0%). The major contributors to the fish landings were Indian mackerel 3.28 lakh tonnes (9.39%), Oil sardine 2.51 lakh tonnes (7.20%), Ribbon fishes 2.27 lakh tonnes (6.49%), Cephalopods 2.06 lakh tonnes (5.89%), Threadfin breams 1.99 lakh tonnes (5.69%), Lesser sardines 1.69 lakh tonnes (4.85%), Scads 1.63 lakh tonnes (4.68%), Penaeid prawns 1.63 lakh tonnes (4.68%), Anchovies 1.57 lakh tonnes (4.50%) and Non-penaeid prawns 1.47 lakh tonnes (4.21%)[13].

Since marine fishery resources are renewable and not in-exhaustible management and conservation of these resources are very much essential for sustained production from the seas. The various entrepreneurial opportunities in marine capture fisheries sector are fish harvesting, post-harvest fisheries, and marketing.

Fish harvesting

In marine fish harvest segment, entrepreneurship development is possible in the activities related to boat/vessel building and overhauling yards, net manufacturing and repair units, gear manufacturing and repair units, ice plants/ cold storages, sea navigational and communication equipment's and other associated activities.

Fig. 1.8: Boat buiding

Fish processing and packaging

The marine post-harvest fisheries processing involves many diverse activities linked with the various methods of fish utilization after landing. This includes development of products, packaging, storage, transportation, quality control, etc. The entrepreneurship options comprise setting and operation of pre-processing units, processing units for value addition, ice plants, peeling sheds, fish meal and fish oil manufacturing units, fish drying units, fish curing units, fish canning units, fish chilling units, IQF products manufacture, battered and breaded products manufacture, ready to eat products manufacture, vacuum packaging, modified atmospheric packaging (MAP), active packaging, cold storages, quality control laboratories, etc.

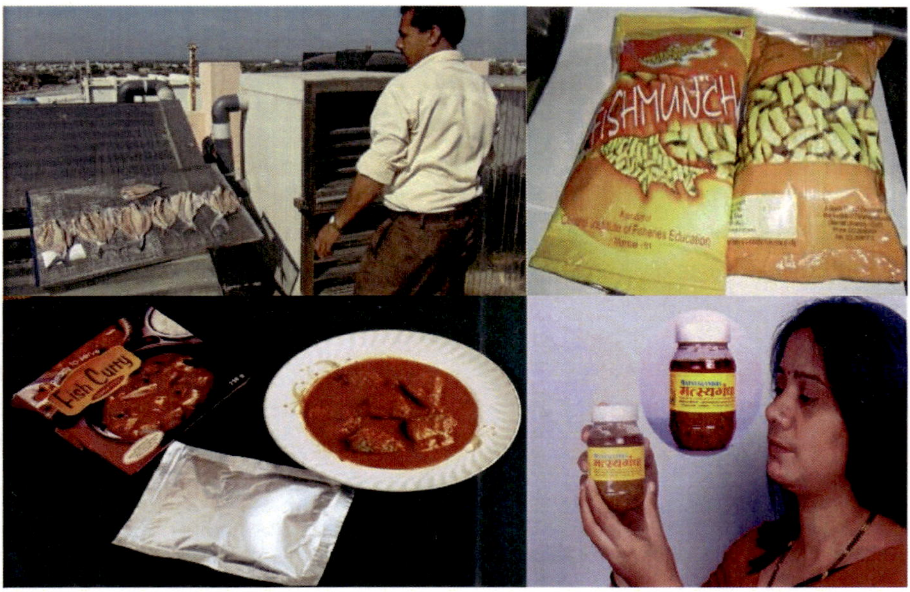

Fig. 1.9: Fish processing and packaging

Fish by-products manufacture

Fish meal, fish silage, fish manure, fish oil, fish protein concentrate, fish protein hydrolysate, fish glue, fish flour, isinglass, pearl essence, fish leather, crustacean waste- chitin and chitosan, pharmaceutical products, neutraceuticals and other bioactive compound extraction from seaweeds, etc.

Fig. 1.10: Herbal products and nutraceuticals from fish

3.3 Fish marketing

Fish marketing also has wide-ranging opportunities to take up an entrepreneurial endeavor. These can be categorized as fish wholesalers, middlemen, fish retailers, fish vendors, fish transporters, cold chain related operators, fish product exporters, online marketing of fresh and processed fish and fish products and many more.

4. Institutional Support for Fisheries Entrepreneurship Development in India

Currently, the Indian fisheries sector is poised for massive growth due to its immense potential for fish production, value addition and as well as due to the various initiatives taken by the government for boosting production in this sector. For entrepreneurship development in the fisheries sector, the various institutional support and government initiatives are described in brief.

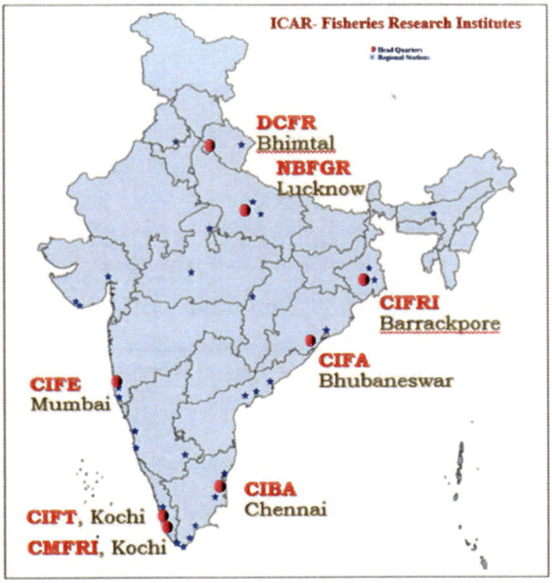

Fig. 1.11: ICAR Fisheries Research Institutes in India

4.1 Indian Council of Agricultural Research

The Indian Council of Agricultural Research is an apex research organization in the field of agriculture and its allied areas. The research and academic programmes in fisheries and aquaculture are coordinated through its eight fisheries research institutes viz., ICAR-Central Marine Fisheries Research Institute (CMFRI), ICAR-Central Inland Fisheries Research Institute (CIFRI), ICAR-Central Institute of Fisheries Technology (CIFT), ICAR-Central Institute of Freshwater Aquaculture (CIFA), ICAR-Central Institute of Brackishwater Aquaculture (CIBA); ICAR-Central Institute of Fisheries Education (CIFE); ICAR-National Bureau of Fish Genetic Resources (NBFGR) and ICAR-Directorate of Coldwater Fisheries Research (DCFR) which are inturn supported by a wide network of their Regional/Research/ Field centres. All the institutes support technology transfer through need-based extension and capacity building of stakeholders through training. Apart from the ICAR institutes there are Fisheries Colleges in the Central Agricultural Universities (CAUs) and State Agricultural Universities (SAUs) which support in entrepreneurship development in the areas of fisheries and aquaculture.

Agri-Business Incubators (ABIs)

To facilitates the entrepreneurship development in the fisheries sector ICAR has setup ABIs in fisheries research institutes namely, ICAR-Central Institute of Freshwater Aquaculture (CIFA), Bhubaneswar; ICAR-Central Institute of Brackishwater Aquaculture (CIBA), Chennai; ICAR-Central Institute of Fisheries Education (CIFE), Mumbai and ICAR- Central Institute of Fisheries Technology (CIFT), Kochi. Through these ABIs, the technologies developed by the institutes are conveyed to the fish farmers, local youths, budding entrepreneurs and other interested individuals. ABIs also assist the potential and existing entrepreneurs for setting up their own business units through providing facilities for R&D, advisory and consultancy services and trainings. ABIs also offer facilities for technology and product development, improvement and upscaling; testing the quality and safety aspects of developed technologies and products; exploring possibilities for commercialization and marketing.

The Government of India has also launched several initiatives for promotion of the fisheries sector. Some of the recent major Government initiatives in the sector are as follows:

4.2 Blue Revolution - Integrated Development and Management of Fisheries (*Neel Kranti Mission*)

The Department of Fisheries, Ministry of Fisheries, Animal Husbandry and Dairying, Government of India in December, 2014 had launched 'Blue Revolution' with a central outlay of Rs. 3000 crores. The Scheme aimed with a focus of an integrated approach for the development and management of fisheries covering both the marine and inland fisheries to ensure a sustained annual growth rate of 6% to 8% in fish production. It also aimed to increase fish production and productivity by utilizing the fisheries resources judiciously, and also doubling income of fishers and fish farmers in five years. The scheme adopted the strategy of encouraging private investment, entrepreneurship development and better leveraging of institutional finance. Skill development and capacity building in fisheries and allied activities and creation of post-harvest and cold chain infrastructure facilities are the channels of achieving the targets. The scheme was implemented from 2015-16 to 2019-20.

The objectives of the scheme were to tap the total fish potential of India on both inland as well as in the marine sector and to triple the production by the year 2020; transforming fisheries sector into a modern industry; ensure active participation of fishers and fish farmers in income enhancement; doubling the income of fishers; and developing nutritional and food security of the nation.

The Blue Revolution resulted in the development of fisheries sector and thus contributing towards the food and nutritional security. Under Inland Fisheries and Aquaculture 22,607 ha area has been developed for pisciculture; established 11 Brood banks for development of species such as Mahaseer, *Channa Striatus* (Murrel), Jayanti Rohu, IMC, Amur Carp, Pangasius, trout, etc.; 535 fish seed hatcheries were established for species such as IMC, trout, Pangasius, Murrel, etc. and 11 hatcheries were renovated / upgraded. 3,679 ha were developed for fish seed rearing units.

Promoted activities for enhancement of income and livelihood opportunities to communities' dependent on wetland through stocking of fish fingerlings in 2,875 ha and also facilitated Integrated Reservoir fisheries development program in 29 reservoirs.

In Marine Fisheries 6 Nos. of fishing harbours were established, 2 Nos existing harbours were modernized /upgraded. Under post-Harvest infrastructure - cold chain has been created through establishment of 221 units of Ice plants, 8 units of Cold storage, 104 units of ice plants cum cold storage, 128 units of renovation of existing ice plants, 303 units of refrigerated & insulated trucks 10 tonnes and one unit of refrigerated & insulated truck 6 tonnes. To provide fish in a hygienic means and to earn better income by the fish vendor, supported Fish Transport & Marketing facilities for procuring 832 units of auto rickshaws, 5614 units of motor cycle with ice box and 225 units of mobile/ retail fish outlets (Kiosk) were set up. Through the Blue Revolution Scheme the fisheries and aquaculture production contribution has increased 1% and 5% to India's GDP and Agricultural GDP respectively[14].

4.3 Fisheries and Aquaculture Infrastructure Development Fund (FIDF)

During 2018-19, the Department of Fisheries, Ministry of Fisheries, Animal Husbandry & Dairying, Government of India created a dedicated fund, FIDF, with a total funds size of Rs 7522.48 crore. FIDF envisages creation of fisheries infrastructure facilities both in marine and inland fisheries sectors and augment the fish production to achieve the target of 15 million tonnes by 2020 set under the Blue Revolution. Besides, the FIDF aims to achieve a sustainable growth of 8-9 per cent, in a move to augment the country's fish production to the level of about 20 million tonnes by 2022-23.

FIDF area of operation includes- Creation and modernization of capture & culture fisheries infrastructure; Creation of Marine Aquaculture Infrastructure; Creation and modernization of Inland Fisheries Infrastructure; Reduce post-harvest losses and improve domestic marketing facilities through infrastructure support; and to bridge the resource gap and facilitate completion of ongoing infrastructure projects.

FIDF provides concessional finance to the Eligible Entities (EEs), including State Governments/ Union Territories and State entities for development of identified fisheries infrastructure facilities through Nodal Loaning Entities (NLEs) namely, National Bank for Agriculture and Rural Development (NABARD), National Cooperatives Development Corporation (NCDC) and All scheduled Banks. Under the FIDF, the Department of Fisheries provides interest subvention up to 3% per annum for providing the concessional finance by the NLEs at the interest rate not lower than 5% per annum. Loan lending period under FIDF is five years from 2018-19 to 2022-23 and maximum repayment period of 12 years inclusive of moratorium of 2 years on repayment of principal.

National Fisheries Development Board (NFDB), Hyderabad is the Nodal Implementing Agency for FIDF Scheme. NFDB has developed an online FIDF Portal. The portal encompasses information vv.r.t FIDF for all stakeholders, online application form for all types of entities, online processing and approval of FIDF applications. Each of the stakeholders have been given separate login for processing of the respective application. The portal also has FIDF Dashboard to monitor the progress of applications on a real time basis. The portal can be accessed using the URL: www.fidf.in[15].

4.4 Fisheries Kisan Credit Cards (KCC)

The Government of India, in the year 2018-19, extended KCC facility to fisheries and animal husbandry farmers to help them to meet their working capital requirements. The detailed guidelines were issued by the Reserve Bank of India (RBI) on 4thFebruary, 2019 covering the eligibility criterion, scale of finance etc. Bank authorities have been instructed to issue KCC within 14 days of receipt of the completed application from the fish farmers.

- For the existing KCC holders the benefits of interest subvention and prompt repayment incentive will be admissible up to the credit limit of Rs. 3 lakhs including fisheries activities.
- In the case of new card holders, the credit limit is Rs. 2 lakhs to meet their working capital requirements for fisheries activities.
- In the KCC scheme @7% is the lending rate to farmers including @2% interest subvention per annum by Govt. of India. Also, another @3% per annum is provided in case of prompt repayment as an additional incentive as per the existing guidelines.
- This implies that the farmers repaying promptly as above would get a loan @ 4% per annum effectively for loan amount up to Rs 2 lakhs[16].

4.5 Pradhan Mantri Matsya Sampada Yojana

The Department of Fisheries, Ministry of Fisheries, Animal Husbandry and Dairying, Government of India is implementing a flagship scheme "Pradhan Mantri Matsya Sampada Yojana (PMMSY)", a scheme to bring about Blue Revolution through sustainable and responsible development of fisheries sector in India with an investment of Rs 20,050 crore for a period of five years, in effect from the financial year 2020-21 to the financial year 2024-25. The scheme intends to address critical gaps in fish production and productivity, quality, technology, post-harvest infrastructure and management, modernisation and strengthening of value chain, traceability, establishing a robust fisheries management framework and fishers' welfare.

The PMMSY is an umbrella scheme with two separate Components namely (a) Central Sector Scheme (CS) and (b) Centrally Sponsored Scheme (CSS). The Centrally Sponsored Scheme (CSS) Component is further segregated into Non- beneficiary oriented and beneficiary orientated sub-components/activities under the following three broad heads: (i) Enhancement of Production and Productivity (ii) Infrastructure and Post-harvest Management (iii) Fisheries Management and Regulatory Framework. It would focus on all round development of Fisheries sector through a bunch of diverse interventions along the fisheries value chain right from production to consumption.

The scheme would create a conducive environment for private sector participation, development of entrepreneurship, business models, promotion of ease of doing business, innovations and innovative project activities including start-ups, incubators etc. in fisheries sector.

Suitable linkages and convergence will be fostered with various centre and state government schemes such as "Sagarmala Project" of the Ministry of Shipping, PMKSY of Ministry of Food Processing Industries, MGNREGS and National Rural Livelihoods Mission of Department of Rural Development, Rashtriya Krishi Vikas Yojana (RKVY) and others for consolidating and amplifying outcomes.

Adequate allocations will be made for infusing new and emerging technologies like Recirculatory Aquaculture Systems, Biofloc, Aquaponics, Cage Cultivation, etc. to enhance production and productivity. Thrust will be given on species diversification, promotion of high value species like Seabass, Scampi, mud crab, and *Penaeus indius* (shrimp) including establishing a national network of Brood Banks for all commercially important species. And genetic improvement of commercially important fish species and establishing Nucleus

Breeding Centre for self-reliance in shrimp brood stock. Also addressing diseases, antibiotics and residues issues for aquatic health management.

Insurance coverage for fishing vessels would be introduced. Youth would be engaged in fisheries extension by creation of 3347 Sagar Mitras in coastal fisher villages. Besides, large number of Fisheries Extension Services Centres would be set up in private space to create job opportunities to young professionals. State of the art whole sale fish markets, retail markets. E-marketing and E-trading of Fish- A National platform would be created.

With the introduction of PMMSY the government is targeting by 2024-25 to raise fish production to 22 million metric tons, increase in the contribution of GVA of fisheries sector to the Agriculture GVA to about 9%, enhancement of productivity to about 5 tonnes per hectare, reduction of post-harvest losses to about 10%, double export earnings, generation of direct and indirect gainful employment opportunities, enhancement of the domestic fish consumption and doubling of incomes of fishers and fish farmers. Along with encouragement of private investment and facilitation of growth of entrepreneurship in the fisheries sector[17].

4.6 National Fisheries Policy 2020

National Fisheries Policy was drafted by the Department of Fisheries, Ministry of Fisheries, Animal Husbandry and Dairying, Government of India in September 2020 aiming to integrate the various existing policies related to fisheries in India. The National Fisheries Policy 2020 has been drafted merging three existing policies, namely: (1) National Policy on Marine Fisheries, 2017 (NPMF); (2) Draft National Inland Fisheries and Aquaculture Policy (NIFAP) and (3) Draft National Mariculture Policy (NMP). Also, Post-harvest elements have also been integrated into the draft policy. The Policy will be released after consultations with all stake holders.

The National Fisheries Policy 2020 aims to develop, harness, manage and regulate capture and culture fisheries in a responsible and sustainable manner. The Policy will ensure a productive integration with other economic sectors, such as agriculture, coastal area development and eco-tourism, to meet the goals of the 'Blue Economy'.

The vision of the National Fisheries Policy 2020 is to develop an ecologically healthy, economically viable and socially inclusive fisheries sector. Another vision of the policy is to provide food & nutritional security to the country in a sustainable manner. The Policy's mission is to develop, manage, regulate and conserve responsibly the fisheries resources[18].

4.7 Indian Marine Fisheries Bill, 2021

This is a Bill envisioned to promote the livelihood and socio-economic well-being of traditional and small-scale fishers, to provide for the sustainable development of fisheries resources in the exclusive economic zone of India and to ensure responsible harnessing of fisheries in the high seas by the Indian fishing vessels and for matters connected therewith or incidental thereto. (This Act may be called the Indian Marine Fisheries Act, 2021. It shall come into force on such date as the Central Government may notify in the Official Gazette)[19].

4.8 Government program to encourage fisheries entrepreneurs

The Department of Fisheries, in association with Startup India, inaugurated the "Fisheries Startup Grand Challenge" on 13th January 2022. The challenge was launched with an objective to provide a platform to startups to showcase their innovative solutions within the Fisheries and Aquaculture sector. The challenge was expected to foster startup culture within the sector and to establish a strong foundation of the entrepreneurial model, the Department of Fisheries has earmarked funds to the tune of Rs. 3.44 Crore for the challenge. The challenge received applications from 121 startups across the problem statements. After a rigorous analysis, 12 startups have been selected as winners of the challenge. The selected 12 winners of the challenge under the Pradhan Mantri Matsya Sampada Yojana were awarded a cash grant of INR 2.00 Lakh each to help them continue building their innovations on the occasion of National Fish Farmer's Day on 10th July, 2023 at Mahabalipuram, Tamil Nadu. Also, the National Fish Farmer's Day celebration event at Mahabalipuram on 10th July, 2023 featured an exhibition of 30 exceptional startups from across the country, showcasing the best amongst fisheries innovators in the country. Curated sessions were also arranged for the startups to help them grow their business further[20 and 21].

5. Way Forward

The fisheries sector is expected to generate better momentum with the implementation of PMMSY, FIDF, Fisheries KCC, National Fisheries Policy, 2020 and Indian Marine Fisheries Bill, 2021. The considerations of food security, livelihood generation and entrepreneurship opportunities are in forefront under these schemes. Also, the strategies and management practices proposed in these initiatives will help in providing not only livelihoods and affordable, nutrient dense food for those who need it most but also gateways to take up commercial ventures.

On the occasion of on India's 76th Independence Day, the Prime Minister of India, Shri Narendra Modi proposed the period of coming 25 years up to the 100th year of India's Independence as Amrit Kaal[22]. In the coming 25 years, during the period of Amrit Kaal it is envisaged to streamline all the aspects of country's growth and advance the nation into a developed country. India is expected to achieve its ambitious set targets through India@2047- a vision plan for a 'future ready India' that befits the 100th year of Indian Independence. Preparation of Action Plan and Document of Vision India@2047 is underway by NITI Aayog. Vision India@2047 will be guided by the aspiration for India@2047, like attaining new heights of prosperity; making best facilities available both in villages and cities; and building world's most modern infrastructure. Vision@2047 will reflect various issues such as global value chains, leveraging sunrise sectors growth, climate change, R&D, innovation & entrepreneurship, employment issues, human capital development, governance, and many others to give a vision for a new developed India.

6. Acknowledgements

The authors gratefully acknowledge all the ICAR- Fisheries Research Institutes for the photographs.

References

Azadi ka Amrit Mahotsav. Available at: https://amritmahotsav.nic.in/about.htm (accessed on 04 December 2023)

Chand, R. and Singh, J. 2023. From Green Revolution to Amrit Kaal- Lessons and Way Forward for Indian Agriculture. NITI Working Paper 02/2023. Available at: https://niti.gov.in/sites/default/files/2023-07/Aggricultrue_Amritkal.pdf (accessed on 04 December 2023)

CSS - Blue Revolution (Neel Kranti Mission) Integrated Development and Management of Fisheries. Available at: https://nfdb.gov.in/welcome/blue_revolution and https://www.nfdb.gov.in/PDF/Blue%20Revolution%20Final.pdf (accessed on 04 December 2023).

Draft Compendium of Startup-Specific Initiatives. Available at: https://www.startupindia.gov.in/content/dam/invest-india/Templates/public/Draft%20of%20Compendium%20of%20Startup%20Specific%20Initiatives%20(1).pdf (accessed on 04 December 2023).

Fisheries KCC. Available at: https://dof.gov.in/fisherieskcc (accessed on 04 December 2023)

Guidelines for Fisheries and Aquaculture Infrastructure Development Fund (FIDF). Available at: https://nfdb.gov.in/PDF/FIDF_Final.pdf (accessed on 04 December 2023).

Handbook on Fisheries Statistic (2022) Department of Fisheries, Ministry of Fisheries, Animal Husbandry and Dairying, Government of India, New Delhi. Available at: https://dof.gov.in/sites/default/files/2023-01/HandbookFisheriesStatistics19012023.pdf (accessed on 04 December 2023).

https://caa.gov.in/uploaded/doc/LIST_OF_REGISTERED_HATCHERIES_314.pdf (accessed on 04 December 2023).

https://caa.gov.in/uploaded/doc/LIST_OF_REGISTERED_NAUPLII%20REARING_HATCHERIES.pdf (accessed on 04 December 2023).

https://pib.gov.in/PressReleaseIframePage.aspx?PRID=1938122

https://pib.gov.in/PressReleasePage.aspx?PRID=1654543 (accessed on 04 December 2023).

https://pib.gov.in/PressReleasePage.aspx?PRID=1852024 (accessed on 04 December 2023).

https://pib.gov.in/PressReleasePage.aspx?PRID=1962015 (accessed on 04 December 2023).

https://www.pib.gov.in/PressReleasePage.aspx?PRID=1909213 (accessed on 04 December 2023).

Jacob, C.T., Sugunan, V.V., Meenakumari, B. and Mandal, R. (2019) Mainstreaming Biodiversity: Inland Fisheries and Aquaculture - A key for food and nutritional Security, Centre for Biodiversity Policy and Law, National Biodiversity Authority, Chennai, India. Available at: http://nbaindia.org/cebpol/pub/mainstreaminginland.pdf (accessed on 04 December 2023).

Jacob, C.T., Suresh, V.R., Sugunan, V.V. and Meenakumari, B. (2020) Conservation of aquatic biodiversity for achieving India's food, nutritional and livelihood security with special reference to inland fisheries and aquaculture. Journal of Fisheries Research 4(1), 30-35. Available at: https://www.alliedacademies.org/articles/conservation-of-aquatic-biodiversity-for-achieving-indias-food-nutritional-and-livelihood-security-with-special-referenc.pdf (accessed on 04 December 2023).

Jena, J.K. and Gopalakrishna, A. (2012) Aquatic biodiversity management in India. Proceedings of the National Academy of Science, India Section B: Biological Sciences 82(S2), 363–379. DOI 10.1007/s40011-012-0108-z

Marine Fish Landings in India-2022. Technical Report, CMFRI Booklet Series No. 31/2023. ICAR-Central Marine Fisheries Research Institute, Kochi. Marine Fish Landings in India 2022. Available at: http://eprints.cmfri.org.in/16883/1/Marine%20Fish%20Landings%20in%20India_2022.pdf (accessed on 04 December 2023).

Marine Products Exports. Available at: https://mpeda.gov.in/?page_id=438 (accessed on 04 December 2023).

Pradhan Mantri Matsya Sampada Yojana. Available at: https://nfdb.gov.in/welcome/PMMSY (accessed on 04 December 2023).

Rasal, A., Patnaik, M., Murmu, K., Mohanty, R.K., Sundaray, J.K., Swain, J.K. and Mahapatra, K.D. 2023. Climate smart carp hatchery: Bringing resilience to sustainable fish seed production. Aquaculture (571) 739476. https://doi.org/10.1016/j.aquaculture.2023.739476. Available at: https://www. sciencedirect.com/science/article/pii/S0044848623002508?via%3Dihub (accessed on 04 December 2023).

The Indian Marine Fisheries Bill, 2021. Available at: https://dof.gov.in/sites/default/files/2021-10/Draft_Indian_Marine_Fisheries_Bill_2021.pdf (accessed on 04 December 2023).

2

Agribusiness and Aquapreneurship

Shakti Ranjan Panigrahy

Assistant Professor, International Agribusiness Management Institute AAU, Anand, Gujarat

Abstract

This chapter provides an overview of opportunities for aquapreneurs in agribusiness and aquapreneurship. It also highlights success stories of some entrepreneurs, in fisheries and aquaculture, in India.

Keywords: *Agribusiness, Aquaculture, Aquapreneurship opportunities, Fisheries, Ornamental fisheries*

Introduction

Where there is water, there is fish. India is country accounts 4.2 percent of global water and has competitive advantage in fisheries production. But, unfortunately the sector has been upholding the tag line of fishy (something unpleasant) in spite of its importance in food, income, society and health[1]. Even, the livelihood of backward communities near to the seashore and other water basins cannot be ignored and could be strengthened by enriching their life through fisheries sector development who will be a barricade for the intruders in future. Panigrahy & Nayak (2023) highlighted that sustainable Development Goal also emphasized upon the importance of fisheries in various segments like no poverty (SD 1), zero hunger (SD 2), good health and wellbeing (SD 3), gender equity (SD 5), responsible consumption and production (SD 12), and climate action (SD 13). Day by day the country has also improved its export share both in volume and value through fisheries that stimulate employment, investment, consumption, and future income of the people who integrate them with fisheries. However, fisheries in our country have been facing the tragedy of commons, conflicts, and loss of biodiversity due to target fishing in absence of effective market intelligence on which the complete business ecosystem of the stakeholders are revolving round through day and night.

---

1 FISH (Fish, Income, Society, Health)

Food and Agriculture Organization (FAO) in its report highlighted that marine fisheries consumption suits the developed country whilst inland fisheries for developing and underdeveloped one according to the consumer surplus and disposal income is concerned (FAO,2014). Even growing consumer surplus also triggers protein consumption over carbohydrate that stimulates more fish consumption in future in our country as well. Study says, fish can be taken with rice as better lysine content of former compensate in diet of the consumer. In this juncture, fish as a functional food has an immense importance across the country where marine fisheries target to Southeast Asia, China, European country, Japan, and US and Inland fisheries find its customers in the national sphere[2] (Panigrahy, Vahoniya & Kumar, 2017). In contrast to global average, where per capita fish consumption is around 20kg per year our per capita fish consumption is 9.8kg with target of 12kg by the Indian council of Medical Research (ICMR). No doubt, country is surrounded from three sides (8118km) with sea water, our marine catch is almost stagnant due to infrastructural bottleneck, irresponsible fisheries, climate change and many more. Once we like to improve the fisheries environment and consumption, we have to focus more on inland fisheries; particularly from aquaculture and for that sake the involvement of aquapreneurs with astute business acumen is need of the hour that require the behavioral transformation of self-sustained production practices to commercialized one with paramount risk and return trade off in the changing dynamics of business ecosystem.

No business exists in a vacuum, always impacted by both external and internal forces. Once agriculture turns to agribusiness due to an increase in marketed surplus, many more stakeholders and their activities intermingle with the flow of produce along with reverse flow of fund and information. Fisheries are perishable commodities, need immediate disposal with proper cold chain across the supply chain. Typical nature of fisheries due to its origin and biodiversity, it is always observed a long distance food miles[3] for it from the origin point to consumption. That makes the system cumbersome, challenging and critical for the fisheries and the people who are involved in them.

Agribusiness is that discipline and subject area in agriculture and food science that connects input sector with production; and then production to processing,

[2]Only, *Litopenaeus vannamei* species of shrimp from brackish water has a good share in export basket

[3]As per the statement of Eldhose Mathew, founder of Daily fish (Baby Marine Sea Food Retail Pvt. Ltd.) an e-commerce company, Seer fish travails 1427 Kms from Mumbai, Mackerel travails 426 Kms from Mangalore, Anchovies travails 323 Kms from Tuticorin, Shrimps travail 449 Kms from Mandapam, Tuna travails 1480 Kms from Visakhapatnam and White pompret travails 2305 Kms from Veraval to reach to Kochi.

marketing and trade (Panigrahy, 2023). Like other sectors, agribusiness also has many implications in the fisheries sector development and growth. Once, no one believed that fisheries would be exhausted from the water. However, growing demand of fisheries depleted the resources at one end and developed subsequent linkages with various actors of supply chain processes for better value creation. In this juncture, changing dynamics of agribusiness leads to market integration (both horizontal and vertical integration) of the fisheries in which other functional areas like finance, human resource and operations support the system for the stakeholders; more particularly aquapreneurs. Aquapreneurs are those human resources who always find the opportunities in the fisheries value chain from the source of production to consumption by doing risk-return trade off in an efficient way.

With this back drop, the present chapter is categorized in to following segments viz.,

- Opportunities of aquapreneurs in Agribusiness and fisheries sector
- Risk- Return Trade Off and Fisheries
- Challenges for the Aquapreneurs and Strategy
- Successful Aquapreneurs and Models

Opportunities of aquapreneurs in Agribusiness and fisheries sector

For encashing opportunities from the fisheries sector, we must understand ecological and economic environment of both marine and inland resources. In India, we must catch more than 70 percent of marine fisheries in the month of September and trade them to Southeast Asia and China mostly due to their flexible in quality screening practices. The catching percentage is quite higher in Gujarat state which is a vegetarian in culture followed by lack of infrastructure (processing industries, cold storages) make exporters to sell them to the distance countries without keeping them for future value creation. It was observed that we are getting revenue due to a higher volume of exports than the value from the fisheries. This is the area in which aquapreneurs can start their venture by establishing infrastructures near to the production areas and can add value to the produce. If this can happen, we can target better international market like USA, European Union etc for our catching fisheries[4]. From our observation we understood that fishermen of Veraval, Porbander, Mangrol areas of Gujarat prefer procurement of fisheries through companies

[4]This time China and South East Asian country import fisheries from India, do value addition and sell them to the profitable market like USA and European Union.

than Farmers producer organization as carrying capacity and expertise of former is better over than any sorts of collectivization practices. Even, the time comes now to relook our national market instead of international one by doing effective market segmentation of the consumers accordance to their demographic, psychographic, behavioral and regional centric variables and subsequent consumption.

Another area of opportunities in sea fisheries is observed through in depth catching practices than the distance one. We all know that our major fish catch from pelagic and demersal zone of marine one that creates a situation of more mechanized systems are running after few fisheries which escalates subsidies, labour cost and other hidden cost like informal credit of the auctioneers. Again, this is an area where we can sensitize our aquapreneurs to try their ways in dipper catch of fisheries through infrastructural development, imbibe the culture of thrift through microfinance, grading and sorting of commodities after aggregation of fisheries and target them to their ultimate market.

Another challenges, we faced in marine catch is to target small and less demand fisheries which either sell at very less price or put them again in water leads to eutrophication, deoxygenation, and further movement of fisheries from our exclusive economic zone of 202 nautical miles from the seashore. It is an area of opportunities in which aquapreneurs can address the issues through effective market intelligence in which they can catch target fisheries as per the requirement of the consumers.

As observed, marine fish catch is dwindling around 3 to 4 million metric tonne and shared approximately 30 percent of the total Indian fisheries; marine aquaculture comes as another area to attract budding entrepreneurs in this field. Putting more value to the marine resources, intercropping, and integrated fish farming practices are observed in which rice-fish culture[5], poultry-fish-vegetable culture, shrimp farming are done that will be a profitable for the aquapreneurs. Marine fish production through cage culture helps the entrepreneurs to harvest two crops in a year with better productivity. Integrate Multi tropic Aquaculture is an area of venture marine fish production is going on with seaweed, algae (*S. latissima*) and mussel farming create a niche area of business. These types of practices add value to the aquaculture system due to improve efficiency by utilizing waste and underutilized feed for the growing of species of lower level tropic like plants and invertebrates. Seaweed farming opens the gate of

[5]Traditionally practiced bheries in West Bengal and Pokkali in Kerala are the rice-fish culture methods in which excreta and biodegradable substances of prawns and fish use for rice production through intercropping method

agar, algin, carrageenan, manure, biofertiliser, fodder, bioactive metabolites like niche market for the aquapreneurs. Even, aquapreneurs should also relook other venture of sea weed farming like seed banks of seaweed, processing of seaweed, and pharmaceutical industries for reaping more benefits in the future. Marine nonliving resources are now an area of priority of the Government to generate oil and gas for human use.

Brackish water is another area of opportunities for the aquapreneurs where they can produce, market and generate profit from the fisheries; more particularly pearl spot, milk fish, mullet, sea bass, shrimps and prawns. The interventions in this resource add value in the waste land to a productive one by creating employment opportunities to those who are directly or indirectly involve in the brackish water aquaculture[6]. Once an area is developed through shrimp farming (majorly dominating inland fish in India), it brings opportunity for other aquapreneurs who relate to shrimp hatchery, feed companies, processing plants, aqua labs, pharmaceutical companies, and registered exporters in this domain. Besides that, other entrepreneurs who are indirectly support aquapreneurs through their supportive business models like ice factory, pump selling and repairing, grocery, tea stalls, fresh water tank, wooden merchant, shrimp seed carriers, tractors for digging the ponds and many more[7] are benefitted in an adverse circumstances and areas. Means to say that brackish water aquaculture is not a food production in itself through fisheries rather it has many amplifier effects across the country through the people who live in fisheries.

Unlike marine and brackish water fisheries, Fresh water fisheries travel comparatively less distance from the point of production to consumption might be due to infrastructural bottleneck and ill-equipped market yard. Fresh water fisheries like Rohu, Catla, Mrigal, and other species target majorly to the national market. Even producers face the difficulties of getting quality feed and seed at affordable rate in their production practices. High wage rate, increased lease rental, multipurpose use of water sources, and unscientific fisheries production make the fresh fisheries production an inefficient enterprise. This is an area where aquapreneurs may step into in different division of the supply chain from pre harvest to harvest and subsequent post-harvest market disposal. Highly unorganized dominated by cartelization, this freshwater fisheries system needs supportive hand with effective interventions in coming years. Like marine and brackish water aquaculture, freshwater fisheries also need

---

[6]It was seen agriculture provide in an around 200 man-days employment in a year where as shrimp farming generates 700-750 man-days per year from a salt affected waste land water resources.

[7]Highlighted business avenues are called as ancillary units that always support main aquaculture for flourish.

the support of an ancillary system in which many aquapreneurs may look their venture stability and profit. As fisheries are voluminous, bulky, perishable, and unstandardized, aquapreneurs may bring paramount solutions[8] for the upliftment and up gradation of this sector.

Many examples and cases witnessed about successive models of fish disposal to the consumers despite many difficulties and deficiency in the fisheries sectors. Aquapreneurs take advantage of floor space in Spencer like super market through vendor management. E-Choupal Sagar is the successive venture of ITC for collecting of fish from the fisherman and sell through their outset and branding (Kumaran et al., 2010). Previously, loads of freshwater fisheries were marketed through trucks from East Godavari, Andhra Pradesh to the Howrah, West Bengal which are graded, sorted, packed, and marketed through trucks by the aquapreneurs in which post-harvest losses minimized and accountability of the supply chain process has been improved. Through market research, now a days small fish is targeted to the tribal pockets and joint families where as large fishers are marketed to nuclear family and urban sector. Many aquapreneurs ventured into the market by the policy of price discrimination. They segmented the market according to income homogeneity (marine fish for higher income group), peak demand criteria (Sunday and Wednesday in Anand District, Gujarat), species (Rohu, Catla for Bangalies, Pungacious varieties for hotels for chilli fish), and through online for the consumer who avoid unhygienic market space. Aquapreneurs also categorized the markets as per the live fish, frozen fish, fresh fish, dry fish, de-skinning fish etc and set the price for their product.

Ornamental fisheries at the shelf of the house not only make the environment lively but also a lucrative venture for the aquapreneurs in 21st century. The business ecosystem of the ornamental fisheries touches around 555cr and could be touched to 5000cr after the support of Government through Pradhan Mantri Matsya Sampada Yojna. New budding entrepreneurs are coming to this field despite high logistic cost from the source of origin to point of consumption. It is estimated that presently about 1.25 percent of urban households are keeping an aquarium which could grow to 5 percent during the present decade. It may grow even faster as the country is transforming from a developing to a developed nation. It is noted that the number of aquarium hobbyists is 10-20 percent of total population in other developed countries of

[8]Aquapreneurs may find opportunities in grading, sorting, icing, generating insulated materials (saw dust), thermocol sheets and even through hygienic retail outlet for disposal of fish to the consumers.

the world. It was even attempted to set freshwater aquarium in different states to rationalize demand of the consumers. Again, growing business through aquarium industries bring other ancillary industries across its supply chain and support the stakeholders in their socioeconomic development.

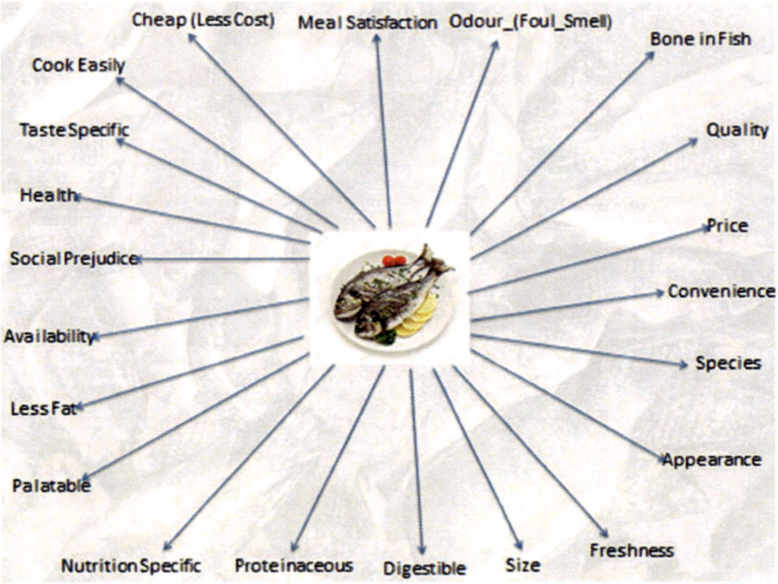

Fig. 2.1: Fish evaluation attributes perceived by the consumers

To access a greater consumer base, now a day aquapreneurs need to adapt technological changes in their domain of business. It helps them to secure their enterprise with greater speed and accountability. Visibility of the product in the era of competition is possible through technological support and traceability. Even, technological applications in business create rational use of resources that leads to more productivity and less wastages. Starting from production of fisheries, an aquapreneurs can take the help of robotics in pond preparation, remove disease fish, feeding, monitoring health, injecting vaccines and many more. Drones application helps in fish monitoring both in and above the surface of water along with data gathering through artificial intelligence (AI) and cloud computing that leads to better aquaculture operations. Presently, aquapreneurs can take the support of sensor or remote sensing to understand fish health, feeding status and feed utilization under the water which leads to improved efficiency in aquaculture. Artificial intelligence makes the system faster and efficient one for the aquapreneurs. Augmented reality facilitates underwater drones and robots and checks net holes and fish mortality. Aquapreneurs can also take advantage of virtual reality in teaching, training, and extension activity. As fisheries sector transform from subsistence to commercial one in the advent of agribusiness and technology, it will create an avenue of opportunities for aquapreneurs in coming days.

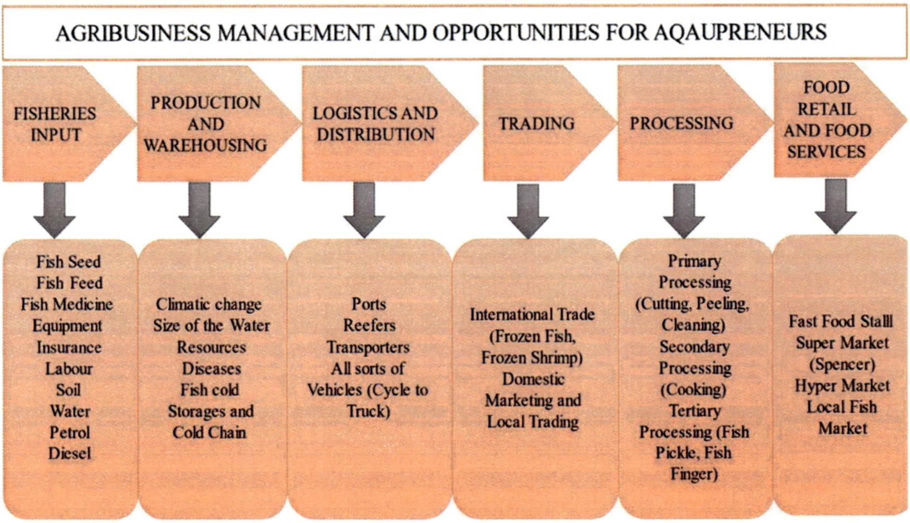

Fig. 2.2: Agribusiness Management and Opportunities for the Aquapreneurs

It is highly epitomized as 'Agribusiness- A bouquet of Entrepreneurship'. Many areas from input sector to output sector are still yet to be untapped in the domain of entrepreneurship. So, resource use efficiency and its optimized utilization can only be possible through value addition across the supply chain that helps towards value creation of the consumers. Market research for each element of the supply chain in agribusiness sector generates opportunities for the aquapreneurs. As fisheries are dwelling under informal sector, new budding aquapreneurs should come to this field for reaping benefits for them and for the society as well.

Risk- Return Trade Off and Fisheries

There are three categories of risks involved in fisheries business viz., financial risk, production risk, and market risk. Financial risk takes in to account of the cost of capital whereas production risk deals with climate, disease, production practices and market risk takes into account low price and its volatility. An aquapreneurs needs to know about risk adheres in their business or they can take the help of aqua clinic and aqua business centers to overcome any challenges. When financial aspect will be considered, it was observed that aquapreneurs takes the required money from different sources like informal (money lender, land lord, family, friends, relatives, traders) and formal one (banks, financial institutions, venture capitalists). Needless to say, informal sources of financing are flexible in process but higher in cost of capital whereas formal sources of lending is just the reverse one. Still, many instances were observed that aquapreneurs depend more on informal sources for money than the formal one.

Once, cost of capital from different resources is calculated, it is essential to produce and market the product in an efficient way. It is desirable to be in business when your profit (net margin) is higher than your cost of capital. An aquapreneurs should understand the investment criteria (fixed asset to current asset[9]) and then to adjust or readjust its financial resources for getting their desired profit.

Table 2.1: Cost of capital bifurcation of fish farmers

Particulars	**Amount**	**Percentage of total investment**	**Ratio of percentage of total investment**	**Cost**	**Actual Cost**
Friends and relatives	250000	50	0.5	0.07	0.035
Money lenders	100000	20	0.2	0.25	0.050
Own fund	150000	30	0.3	0.04	0.012
Total	500000	100	1.0	0.36	0.097

A hypothetical example highlighted in table 1 that about 9.7 percent of cost of capital was embedded in the fishery business and an aquapreneur needs to go for better profit margins to overcome their financial leverage. From the current scenario of inland fish production, it was observed that a producer invested around 70 rupees per ha in one kg fish production and reaped around 38 rupees as a net margin from it. It is around 54 percent of return (0.54) in the business that is quite higher than the investment cost (0.097). From the calculation part, it is obvious that fisheries are a profitable enterprise. However, a great deal of discussion arises due to the perishability nature of the produce and volatility of demand in market. Still, aquapreneurs find fisheries are a profitable business and any support from the government retards the risk and volatility in the enterprises.

[9]Investment on fixed asset is called as capital budgeting and investment of current asset is called as working capital management.

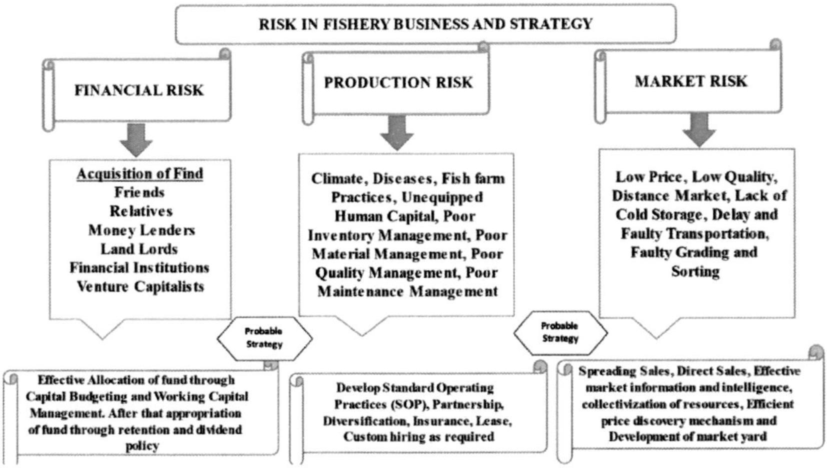

Fig 2.3: Risk in Fisheries Business and Probable Strategy

Challenges for the Aquapreneurs and Strategy

The biggest challenge faced by the aquapreneurs, or any fishery entrepreneurs, is that water is a free resource in which anyone can enter and do fish farming. If that is the thing, fisheries production is always at stake. Responsible fisheries through cooperatives (FPOs) are the solutions that protects fisheries in the water ecosystem. Illegal, Unregulated and Unreported practices are the parts and parcels of the marine ecosystem that needs to be controlled through education, awareness, and sensitive programmes through NGOs and government machineries (Panigrahy & Nayak, 2023). The best way of doing business will only drive more aquapreneurs in this field. As sea water has no ownership right for a long term basis to any one, so no one can like to invest on it for sustained fish production. This is the area in which we must identify more and more aquapreneurs by providing license and sensitize them for more investment on marine aquaculture ecosystem. Even, to boost investment on sea water resources, working capital in the form of money, man and materials are highly needed. Aquapreneurs may provide the solution to the needy one by generating micro financial opportunities at the nearby areas of fishermen.

Climate change is found as an unavoidable element in fish production to supply. Due to climate change, pelagic fisheries are affected a lot and some of the species like Indian Mackerels go deeper into the sea. While tuna fisheries migrate to other countries, oil sardines open their boundaries from one state to another. To counter act this hurdle, it is dire essential to give value to each fishery as fish demand impacts on its diversity. Judicious use of fisheries can only be possible through effective market segmentation and emphasize on attributes-based consumption through awareness programmes[10] and policies. As highlighted by Dr. Shyam S Salim, Principal Scientist, ICAR- CMFRI, Kochi, fish market needs to be segmented on the basis of income, locale (urban/semi-urban, rural), price (high value, low value), sales (online, terminal, retail), species (marine fish, inland fish, farm fish), time (season, festival), orientation (religion), and taste & preference. As small and cheap fish is also used as fish industry and beef industry in their feed production, we need to promote domestic consumption of fish by the human being. Otherwise, many more fisheries will be extinct from our own domain of resources either through natural changes (climate) or manmade interventions (industry demands) in coming future. Aquapreneurs have the opportunities to enter the fish consumption market by developing fish-market grid by which untapped demand along with current consumption could be tapped efficiently in the country.

Another major challenge in fisheries sector is that what to grow and how much a market is capable to absorb. Unlike, poultry sector where consumers prefer value added products and different cuisines in the menu list originated from same species broiler, fisheries consumption is a mere species oriented and changes regularly across the days and in different consumption. So, identifying demand is a herculean task. Any oversupply reduces the retail price near the production centres or increases the inefficiency of the market when products a distance place for their consumption. So, a need bases market information system is required. Informal communication of the price and disposal of the fisheries are also done among the producers through cartel network system; means one producer harvest Rohu fish while another harvest catla fish for rationalizing the market demand. Similar issues of market dysfunctionality are also marked in the international trade where exports suit as per the containers' capacity[11]. Any more or less the capacity increases market inefficiency and decreases producers' share in consumer rupees.

Grassroot issues also impact a lot on fish production to consumption. Growing lease value, labour cost, insufficient fingerlings, and quality less feed decrease the production efficiency in the captivated fisheries sources (Nayak et al., 2022). Along with this it was also observed that some fisheries ponds are also used as other house chores activities like bathing, cleaning of utensils, animal washings and many more. An unhygienic way of fisheries production creates a negative positioning among the consumer for the fish. Aquapreneurs should

[10]We should promote fish consumption by highlighting fish as the cheap source of protein, contains omega 3 fatty acid that has immense health benefits and put the product at right place through organized retailing.

[11]20tons capacity-based container suits to those fishers which are harvested maximum. So, shrimp exports more than the marine fisheries as it is difficult to predict the later categories of fisheries which are captured based.

come forward and replicate best management practices in grassroot fisheries production for garnering benefit from it. Few fisheries entrepreneurs do green net barricade, electrification, multiple small size ponds for maintaining water quality, data documentation, and activity wise labour management in freshwater fisheries production.

As we know, economic development and social development moves hand in hand whereas economic development rebuilds social structure and its development. Fisheries sector is highly dominated by backward classes who could not have requisite resources to reap benefits in their credit. In this juncture, it is essential to link the poor human resources with the mainstream of development. Fisheries cooperatives are one of the steps in this regard in which economic upgradation is possible through social upliftment as highlighted in the caselet 1.

Caselet 1:Matshyafed Fish Cooperatives: Story of fisherman from rigs to rashes

Fisherman not only grow fish but also grow with fish in Indian context but their measurable livelihood due to many of the causes from input cost escalation to management practices and further marketing of the produce always need an institutional support like cooperatives to overcome the bottleneck of drudgery from their limitations. The journey of Matshyafed Fish Cooperatives speaks the story of poor man livelihood in this regard where support and technological interventions make them connect to the mainstream of development where economic improvement leads to social empowerment. Once, the fishermen without any infrastructure depend on boat owners for catching of fish where they only got one third of revenue of the marketed fisheries. Boat owners took the rest while fishermen got less incomparable to their risk they have taken. After that with the support of National Fishery Development Board through institutional upgradation in the form of cooperatives, the people connected, and integrated them and capable to catch more fish which lead to an era of value addition, value creation and upbringing of social structure in due course of time. Aquapreneurs should find their venture in the untapped potentiality of the fisheries resources.

Successful Aquapreneurs and Models

Dr. Manoj Sharma, the founder of Mayank Aquaculture Pvt. Ltd, the person who transforms the life of many inhabitants and migrant labour of Surat, once spoke in open platform of pre vibrant Gujarat, Anand Gujarat, "Do not presume fish as a food only, just see its rippling effect on socioeconomic status of people who connect with it". Economics always say that the income has the amplifier effect and here in this successful story it was also felt that imbibed results changed many followers both in fisheries and non-fisheries domain.

Mayank Aquaculture Pvt. Ltd. is the baby of the founder and through the chapter how many more successful aquapreneurs come out with his successful journey has been discussed.

Name: **Manoj Mohanal Sharma**

District : Surat

State : Gujarat

Education : Ph.D. (Fisheries)

Occupation : Fisheries Entrepreneur

Mobile Number : 9824112836

Fig. 2.4: Zhingalala Mobile Restaurant

Surat city of Gujarat once known for its diamond and fabric is now epitomized as hub of Shrimps of western India and placed third position after Andhra Pradesh and West Bengal. From a mere 300 metric tons in 2009, shrimp production has crossed 45,000 metric tons per annum in 2019, generating an annual turnover of Rs 2,000 crore from this region make enthusiastic to Manoj Sharma, president of Gujarat Aqua Feed Dealers' Association (GAFDA) and he said, "there is a huge potential of increasing shrimp farming in Gujarat as there is 3.75 lakh hectare of brackish water land available in Gujarat, of which hardly 2 percent is under cultivation." Again, he added, "about 800 farmers are involved in shrimp farming on about 7,000 hectare of brackish water land in Surat, Bharuch, Navsari and Valsad districts of Gujarat and many more area are yet to tapped and explored. If things are going as per the predictions, shrimp farmers will give direct employment to more than 25,000 people and indirect employment to approximately 30,000 people in coming days[12]. He was very much realistic from the facts that the domestic consumption of shrimps in India is negligible and that about 90 percent of the production is exported to the United States, China, Japan, West Asia, China and Vietnam. After Covid 19, time comes now to relook the internal market of India which behaves very promising for the market.

[12]Dr Manoj Sharma once stated that in one hectare of land, farmers can cultivate 5,000 kilogram of shrimps having international value of $30,000. Even if we consider 10% profit, farmers earn $3,000 in a single crop.

The toddler journey started in 1994 when Dr Sharma stepped into Surat after his graduation in fisheries. That was the time when complete Surat was flooded with water and no single pond was available for shrimp farming. The journey was again difficult for an outsider of 19yrs age without any language command in a different state and a different culture. The academic power, persuasion, leadership quality and hard work metamorphoses a sheer man to an aquapreneurs, Dr Sharma stated in a webinar. He witnessed the era of monodon (tiger prawn) from 1995 to 2009 and vennamai (shrimp) from 2009 onwards and represented him as the ambassador of other entrepreneurs along with his journey. He motivated his followers by saying that shrimp farming is the enterprise that covert waste land to wealth land, unemployment to self employment, leads to reverse migration by developing infrastructure in coastal land, a highly labor-intensive enterprise, and a venture that brings foreign currency in the income basket.

<table>
<tr><td></td><td rowspan="2">Days witnessed, when Mr Jignesh sold cyclov yeast (dried preserved cyclov for nurturing young shrimps) to the producers and earned 2000 rupees in a day, now become first founder and contractor of American based company Zeigler feed (100cr company) plant in the coastal area where shrimp farming has been carried out. He is an engineer, not even an aquapreneurs but could capable to change the fortune through aquapreneurship.</td></tr>
<tr><td>Mr Jignesh Contractor
(Phone Number 9825115393)</td></tr>
<tr><td></td><td rowspan="2">The person who is driving a 1.5cr car was a mere fish catcher or a labor who moved here and there for catching of fish and earned 100 rupees in a day. Now he is cultivating 1000 tons of shrimps from his own farm and has 150cr turn over from his aquapreneurs business.</td></tr>
<tr><td>Mr Rahul Sunil Bhagwadara</td></tr>
</table>

<table>
<tr><td>

Mr Shanshah Khan
Phone Number 9879523961</td><td>Day was there, Mr Shanshah Khan and his brother colle cted shrimps (6 to 8kg) from Manoj Sharma in his initial days of struggle and sold a total quantity of 60-70kg in the Mumbai market. Now, as day moves on, Mr Khan established a Bashir & Washi company having a turnover of 500cr within a span of 20 years.</td></tr>
<tr><td>

Saileshbhai Panchal
Phone number 9825405126</td><td>When you have a pond for shrimp farming, at least you need a pump for water management that brings second generation of aquapreneurs in the form of Ami Engineering, established by Saileshbhai Patel. Now, as shrimp farming finds its momentum, it also creates an opportunity for the aquapreneurs.</td></tr>
<tr><td colspan="2">Other Aquapreneurs evolved from the shrimp evolution started by Manoj Sharma</td></tr>
<tr><td>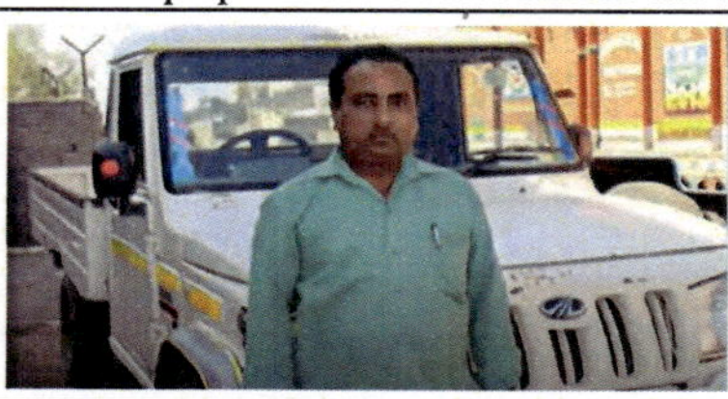
Mr Ramdasbhai Gupta (Driver)
Phone number 9925036842</td><td>Due to growing of shrimp farming, Mr Ramdasbhai purchased three to four carriers that can easily carry 1 tons capacity of shrimps to the distance place. Through the process a journey of an aqaupreneur started from livelihood sustainability to a commercialized business personnel.</td></tr>
<tr><td>
Mr Viraf P Patel (Aqaupreneur of an ice factory) Phone number 9879363230</td><td>Initially, in the year 1994, Mr Patel could able manage his ice factory business only in the period of summer season. Once, shrimp farming started and harvest was done in the month of July, demand of ice increases both for monsoon and winter season as well. Through this process, three generation of Patel family is now in this ice business.</td></tr>
</table>

The branches and offshoot of an aqaupreneur, Dr Manoj Sharma has not been stopped here. In the time of Covid 19, when demand from the international border stopped due to a halt in transport system, another venture was started in the name of Zhingalala Restaurant serving both local and exotic cuisines with the "Pond to Plate" concept. Means to say, aquapreneurship is a journey that is full of excitement, enigma and energy driven path that brings opportunity in the shadow of obstacles.

Concluding Remarks

Aquaculture has the immense potential to transform rural economy by providing opportunities to produce food and profit. Aquapreneurs should grab that opportunities in the era of agribusiness that not only generate resources for the generations to come but also help to utilize resources efficiently towards a sustainable fisheries development. The 21st century is the century of fisheries and fish is going to feed the growing population and, in this way, many lives will be integrated and evolved through fisheries in coming years. Aquapreneurs are that lead in the domain of business that support fisheries and establish socioeconomic development of populace who depend on it.

References

FAO (2014). World Review of Fisheries and Aquaculture. State of World Fisheries and Aquaculture, 1,10

Kumaran, M., Ravisankar,T., Krishnan, M., Vimala,D.D., Mahalakshmi, P., & Ganeshkumar, B. (2010). Unique and innovative cases of emerging domestic fish marketing arrangements in South India. Sustainable Agriculture, XV (1), 18-23

Nayak, A.K., Panigrahy,S.R., Pundir, R.S., & Kumar,V.H.M. (2023). An Economic Evaluation of Freshwater Fish Production and Marketing in Gujarat. Indian Journal of Economics and Development, 19(1), 202-208

Panigrahy, S. R. & Nayak, A (2023). Fishery Cooperatives as a Catalyst for Sustainability. Dana, L.P.,Sharma, N., Kumari,S., & Tripathy, K.K. Cooperatives as a Catalyst for Sustainability: Lessons Learned from Asian Models. Web of Science, Asia-Pacific Business Series: Volume 12

Panigrahy, S. R.; Vahoniya, D.; & Kumar, S. (2017). Production and Consumption of Marine Fish after Globalisation. Indian Journal of Economics and Development, 13(4), 741-746

Panigrahy, S.R. (2023). Opportunity in Agribusiness. Vigyan Varta, 4(7), 57-60

3

Entrepreneurship Opportunities in Fisheries and Aquaculture: NFDB Initiatives

L. Narasimha Murthy

National Fisheries Development Board, Hyderabad -500052

Abstract

The entrepreneurship opportunities in fisheries and aquaculture, specifically focusing on initiatives undertaken by the National Fisheries Development Board (NFDB) in India are summarized in the Chapter. An overview of the current scenario, key areas of entrepreneurial potential, and the NFDB's role in fostering sustainable growth and innovation within the sector are described. Entrepreneurship plays a vital role in the development and sustainability of fisheries and aquaculture sectors. The Chapter also discusses various support programs, funding schemes, capacity-building efforts, and success stories that have emerged through NFDB's initiatives. By showcasing these opportunities and resources the NFDB offers, this chapter encourages aspiring entrepreneurs to venture into fisheries and aquaculture while promoting economic development and conservation.

Keywords: Entrepreneurship Opportunities; Fisheries; Aquaculture; National Fisheries Development Board (NFDB); Sustainable Growth.

Introduction

The Aquaculture and fisheries sector in India is a multi-disciplinary sector that provides food security, employment, economic gains, and socio-economic upliftment for those involved in fish exploitation, production, and processing. Sustainable fishing practices, improvements in fish breeding and husbandry technologies, along with the introduction of new species, have helped increase production levels through small-scale rural aquaculture, largely through governmental and non-government organizations.[1]Aquaculture and fisheries entrepreneurship are crucial for youth economic empowerment in

India, providing and promoting sustainable development and employment opportunities within the fisheries sector. Technological advancements in aquaculture can enhance productivity, optimize resource utilization, minimize environmental impacts, and improve profitability. Aquaculture and fisheries entrepreneurship can promote environmental sustainability by adopting eco-friendly practices and establishing strong market linkages. It has significant potential in rural areas where employment opportunities are limited, reducing migration and contributing to local economic growth. Government initiatives and policies that promote entrepreneurship, youth development, and sustainable fisheries are instrumental in supporting young individuals interested in aquaculture.

Fisheries and aquaculture sector is highly diverse (Figure 1) and an important source of livelihood for about 28 million fishers and fish farmers in the country. India is the 3rd largest fish producing country with about 8% of global fish production. Globally, India stands 2nd in aquaculture production and is one of the top shrimp-producing and seafood-exporting nations. The fisheries sector in India has shown a sustained annual average growth rate of 8.61% for the 8-year period from 2014–15 to 2021–22 (at constant prices). The sector has reached a record fish production of 162.48 lakh tonnes in FY 2021–22 and has immense potential for growth. Export earnings from the sector reached Rs. 63,969.14 crore during 2022–23.[2].

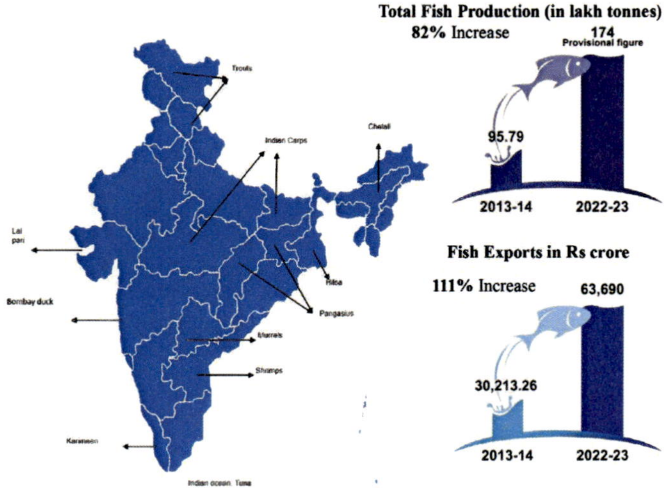

Fig. 3.1: India's Fisheries and Aquaculture Sector

Some of the key initiatives taken up by the Government of India to achive this are represented in Figure 3.2.

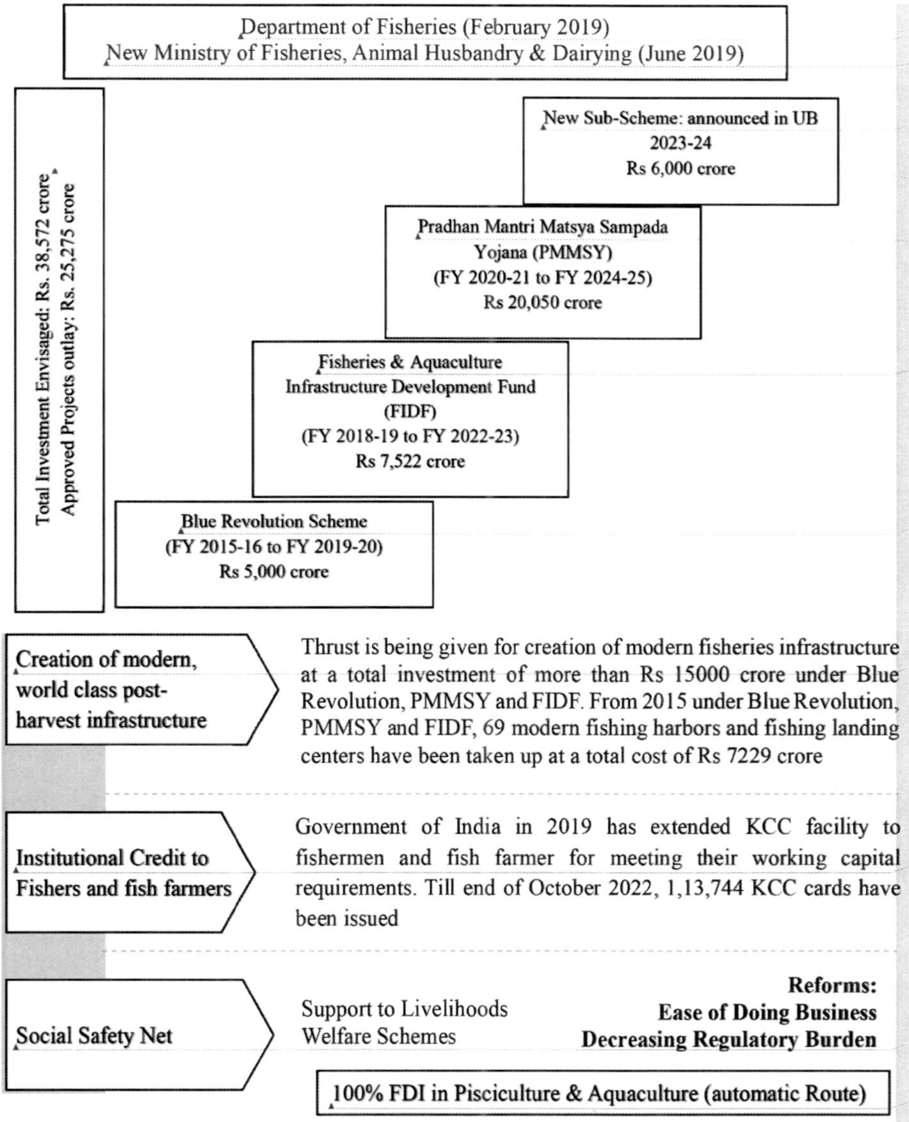

Fig. 3.2: Key initiatives of the Government of India in Fisheries and Aquaculture Sector

The PMMSY (Figure 3.3) is being implemented in the country for sustainable development of the fisheries sector with the objective of enhancing fish production and productivity, developing infrastructure across the fisheries value chain, enhancing domestic fish consumption across the country, and importantly, prioritising the livelihood and welfare of fisher communities. The scheme envisages addressing critical gaps in fish production and productivity, quality, technology, traceability, post-harvest infrastructure and management, creation of a conducive environment for private sector participation, development of entrepreneurship, promotion of ease of doing business, innovations and innovative project activities including start-ups, incubators in fisheries sector. It will further modernize and strengthen the value chain by establishing a robust fisheries management framework and will promote fisher's welfare.[3] Since FY 2020-21 to 2023-24 (August 2023) under PMMSY, a total project cost of Rs. 16,924.02 crore have been sanctioned by the Department of Fisheries, Government of India to implement Centrally Sponsored Schemes (Rs. 15,335. 09 crore) and Central Sectoral Schemes (Rs. 1,588.93 crore) covering 35 states and UTs and other implementing agencies.

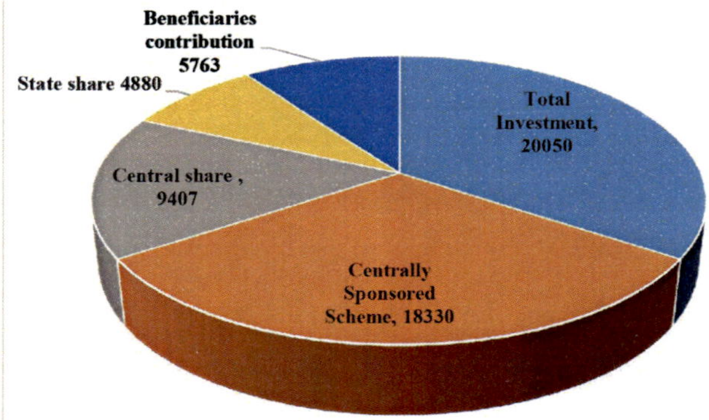

Fig. 3.3: Break-up of PMMSY Investment

Priority action areas for aquaculture transformation

India aims to transform the fisheries and aquaculture sector through policy interventions, Research & Development, promoting sustainable aquaculture practices and strategic partnerships. The four priority action areas in this context are represented graphically in Figure 3.4

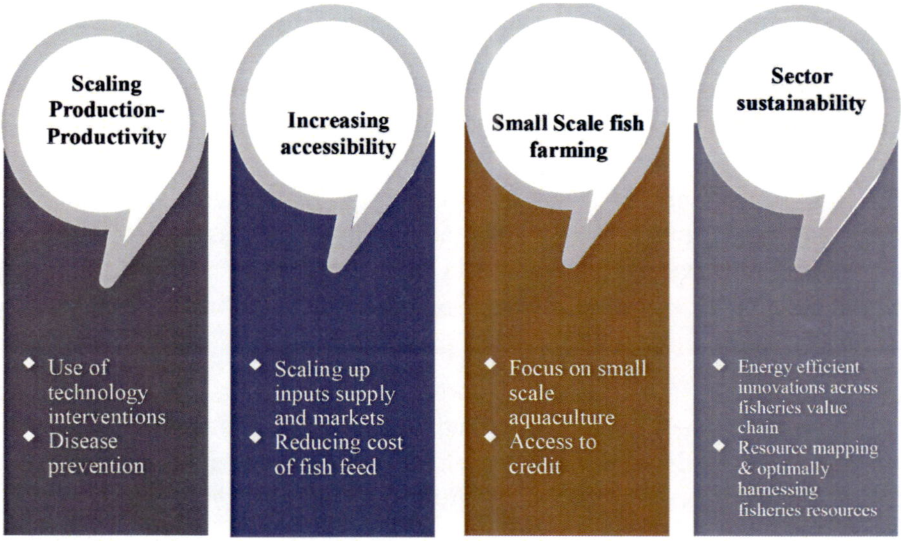

Fig. 3.4: Priority Action Areas for Aquaculture Transformation

Emerging Business Models in Fisheries & Aquaculture Entrepreneurship

The business model for fish entrepreneurship involves identifying a specific niche within the fish industry, such as aquaculture, commercial fishing, seafood processing, or retailing live or processed fish products. Market research is crucial to understanding target customers' preferences and demands and developing effective strategies for product positioning. A unique selling proposition is essential, such as offering sustainably sourced or organic seafood, high-quality products, unique species, or specialized services. Supply chain management is essential to ensuring a consistent supply of fresh fish products at competitive prices. Product diversification caters to different customer preferences and maximize revenue streams. Production facilities and infrastructure should be set up according to the nature of the business, ensuring compliance with safety regulations and obtaining necessary licences or certifications. Sales channels should be chosen, including direct sales at farmers' markets and fisheries co-ops, wholesale distribution to restaurants, grocery stores, caterers, and online platforms. Effective marketing and promotion strategies should be developed, highlighting the unique qualities of the products. Financial management should be robust, including budgeting, cost control, and pricing strategies. Sustainability and regulatory compliance should be prioritized, as should adhering to environmental regulations, responsible sourcing standards, and certifications. Continuous learning and adaptation are essential, with opportunities for professional development through conferences, courses, networking events, and collaborations with research institutions. It's important to note that the specific details of a fish entrepreneurship business model may vary depending on factors such as location, target market, scale of operations, and the chosen segment within the fish industry. Conducting a thorough feasibility study and seeking expert advice from professionals in fisheries management or business development can help refine and tailor your business model to suit your specific goals and local market conditions.

Under PMMSY, a congenial ecosystem for private sector participation, development of entrepreneurship, promotion of ease of doing business, innovations, and innovative project activities, including start-ups, incubators, etc. are encouraged in the fisheries sector. Towards this end, the establishment and handholding of fisheries and aquaculture startups would be a priority intervention under PMMSY. Further, establishment of Fisheries Incubation Centers (FICs) would be supported under PMMSY both through the government and the private sector. They would be managed through the state or central government entities, including the NDFB, and/or through professional private firms or agencies. FICs would provide opportunities For incubators like young professionals and entrepreneurs, fisheries institutes, fisheries researchers, cooperatives and federations, progressive fish farmers, fisheries-based industries, and other entities to showcase their innovations and innovative ideas and technologies in fisheries and commercialize them for the benefit of fishers and fish farmers. This would also help in creating new businesses, entrepreneurs' development (aquapreneurs), and employment opportunities in fisheries. Startups and incubators would be supported under "Innovations and Innovative Projects and Activities, Technology Demonstration, including startups, incubators, and pilot projects" of the Central Sector Component of PMMSY, for which guidelines have been issued in the Operational Guidelines of PMMSY.[7]

Initiatives for Supporting Livelihoods of Fishers and Fish Farmers

1. Innovative Technologies Driving Entrepreneurial Opportunities

a) **Fish Farming:** Opportunities in freshwater and marine fish farming, such as cage culture, pond-based systems, or integrated multi-trophic aquaculture (IMTA).

b) **Hatcheries & Seed Production:** Entrepreneurial prospects in hatchery operations for producing quality seeds of commercially important species.

c) **Aquaponics:** Exploring the integration of fish farming with hydroponic cultivation to create sustainable and efficient food production systems.

d) **Value Addition & Processing:** Opportunities in adding value to fish products through processing, fisheries by-products utilization, and marketing of processed seafood products.

e) **Export-oriented Enterprises:** Identify potential markets for Indian fisheries and aquaculture exports and discuss entrepreneurial opportunities for establishing export-oriented enterprises.

f) **Recirculating Aquaculture Systems (RAS):** Efficient water filtration techniques

g) **Genetic improvement programs:** Selective breeding for desirable traits

2. Market Dynamics Shaping Entrepreneurship Opportunities

a) Growing demand for sustainable seafood products

b) Rising interest in niche markets (e.g., organic or eco-certified products)

c) Value-added processing innovations

3. Policy Frameworks Supporting Entrepreneurial Ventures

a) Government subsidies/grants promoting start-ups

b) Regulatory frameworks ensuring environmental sustainability

c) Supportive policies encouraging innovation

NFDB Initiatives promoting Entrepreneurship

The Fisheries sector with its important role played in the socio-economic development of the country has become a powerful stimulant for the growth of a number of subsidiary small, medium and large-scale industries. To accelerate its development and to translate the research results arising in the field of fisheries and aquaculture sector, National Fisheries Development Board (NFDB), an autonomous body under the administrative control of the Department of Fisheries (DoF), Ministry of Fisheries, Animal Husbandry and Dairying, Government of India (GOI) has taken up several initiatives. NFDB has a pivotal role in implementation of the PMMSY along with the DoF, GOI. It is the implementing agency for 'Entrepreneur Models' under Central Sector Scheme of PMMSY[3,4]. The funding pattern for 'Entrepreneur Models' is given in Table 1. NFDB invites and evaluate the project proposals for Entrepreneur

Models in fisheries and aquaculture under PMMSY and sanctions it to be considered under the Scheme based on the evaluation. Viable entrepreneur models can be under three broad categories;

(1) **Investment driven models:** These will include business models based on a congenial policy environment, with structured incentives and access to water resources to support development of fisheries enterprises.

(2) **Industry driven models:** These include contract farming-based business models that boost engagement between entrepreneurs and fishers & fish farmers along the fisheries value chain (including seaweed, ornamental and recreational fisheries) covering production, processing, post-harvest and value addition.

(3) **Incubator driven models:** To promote technology and value chain-based entrepreneurs through tech-enablement, tech transfer, training, mentoring and handholding.

Table 3.1: Funding Pattern of Entrepreneurship Model

Category	**Government Assistance (maximum limit)**	**Beneficiary contribution**	**Bank Loan (minimum limit)**
General	Up to 25%	Minimum 10% of the total project cost	Up to 65% of the total project cost
SC/ ST/ Women	Up to 30%		Up to 60% of the total project cost
• Ceiling limit of subsidy in case of General category: Rs.1.25 crore per project. • Ceiling limit of subsidy in case of SC/ ST category: Rs.1.50 crore per project. • Funds will be released in 3 instalments as back-end subsidy (20:50:30)			

Apart from this, NFDB would undertake need-based, beneficiary-oriented fisheries development activities to address the priorities of the fisheries sector, which are either unmet or suboptimally or partially met, under the Centrally Sponsored Schemes (CSS) of PMMSY. Here are the areas in which NFDB supports entrepreneurship in the fisheries and aquaculture sectors.:

Financial Support: NFDB provides financial assistance through various schemes and programmes to support entrepreneurial ventures in fisheries and aquaculture. This includes funding for setting up fish farms, hatcheries, processing units, cold storage facilities, and marketing infrastructure.

Skill Development: Recognizing the importance of skill development for successful entrepreneurship, NFDB offers training programmes and capacity-building initiatives. These programmes provide aspiring entrepreneurs with essential knowledge on modern fish farming techniques, aquaculture management practices, post-harvest handling methods, value- added processes, and marketing strategies.

Technology Adoption: To foster innovation and technological advancements, NFDB promotes the adoption of new technologies in fisheries and aquaculture, such as recirculating aquaculture systems, biofloc technology, sustainable feed formulations, and the use of IoT (Internet of Things) devices for monitoring parameters like water quality or feeding patterns. These technologies translate into efficient production systems, reduced environmental impact, and improved productivity, enabling entrepreneurs to undertake profitable ventures.

Market Linkages: The success of any entrepreneurial venture relies on effective market linkages. NFDB initiates programmes that facilitate direct access between producers and consumers by developing value chains, and elaborating with retail chains, fishing cooperatives, and export-oriented companies. They also encourage entrepreneurs to participate in exhibitions, fairs, and trade shows where they can showcase their products, gain exposure, forge business partnerships, or attract potential buyers.

Policy Advocacy: NFDB actively engages with policymakers, industry stakeholders, and regulatory authorities to advocate for favourable policies and regulations that support entrepreneurship in the fisheries and aquaculture sectors. This includes addressing issues related to licencing, permits, land acquisition, credit availability, and export-import procedures.

Information Dissemination: NFDB plays a vital role in disseminating information on entrepreneurial opportunities through its website, pamphlets, brochures, and workshops. They provide comprehensive guidance on various aspects of starting and managing fish farms, hatcheries, fish processing units, cold storage facilities, and other allied businesses.

Networking Opportunities: To foster collaboration and networking among entrepreneurs, NFDB organises conferences, symposiums, workshops, collaborative research projects, and industry-academia interactions. These platforms facilitate knowledge exchange, business partnerships, and access to new markets or technologies.

Conclusion

Entrepreneurship opportunities in fisheries and aquaculture offer immense potential for economic growth, employment generation, rural development, and food security. NFDB's initiatives aim to harness this potential by providing financial support, skill development programmes, technological advancements, and market linkages. Their efforts are instrumental in driving innovation in the sector while ensuring sustainable practices and responsible resource management. Aspiring entrepreneurs can leverage these initiatives to venture into this dynamic field and contribute to the growth of India's fisheries and aquaculture industries.

References

Brugere C. and Williams M. 2017. Women in Aquaculture. Gender in Aquaculture and Fisheries – Expanding the Horizons. https://genderaquafish.org/portfolio/women-in-aquaculture/

DoF. 2020. Pradhan Mantri Matsya Sampada Yojana: Guidelines on Entrepreneur Models in Fisheries and Aquaculture under PMMSY. Department of Fisheries (DoF), Ministry of Fisheries, Animal Husbandry and Dairying, Government of India. https://pmmsy.dof.gov.in/new-download

NFDB. Entrepreneur Model Open-end Advertisement for "Entrepreneur Models in Fisheries & Aquaculture" under Pradhan Mantri Matsya Sampada Yojana dated December 15, 2022. NFDB, Hyderabad. https://nfdb.gov.in/welcome/entremodel

Press Release dated June 28, 2023. Fisheries sector provides livelihood, employment and entrepreneurship to more than 2.8 crore fishers and fish farmers at primary level: Shri Parshottam Rupala. PIB. India. https://pib.gov.in/PressReleasePage.aspx?PRID=1936017

4

Entrepreneurship Opportunities in Dry Fish Sector in India

Amit Ranjan[1] & Siddhnath[2]

[1]Department of Fish Nutrition & Feed Technology, TNJFU-Institute of Fisheries Post Graduate Studies, Vaniyanchavadi, Chennai -603 013
[2]Fish Processing Technology, Guru Angad Dev Veterinary and Animal Sciences University, Ludiana-141 004

Abstract

Dried fish is crucial for nutrition, economy, society, and culture for many of Indian state and play a vital role in nutrition security. In India, dry fishes are widely sold at local markets. The nutritional quality of dried fish remains intact for longer time. Dried fish business can play an important role in the national economy in multiple ways including minimizing post-harvest losses of fish providing a source of animal protein, ensuring livelihoods for rural poor. Dry fish marketing chain begins with a farmer (dry fish producer) and concludes with the final consumer with several intermediaries in between. Dried fish are usually low in price and can be stored and transported for accessibility by the poorest and most vulnerable communities throughout the year.

Keywords: Dry fish, Dry fish producer, fermentation, salting, sidal, smoking, sun drying

Introduction

Dried fish is a staple food for various communities and considered a 'hidden' sub-sector within small-scale fisheries. Dried fish defined broadly as aquatic animals preserved using simple techniques such as sun-drying, salting, fermentation, and smoking that permit storage as foods at ambient temperature for extended periods without specialized packaging (Belton et al., 2022). The nutritional quality of dried fish remains intact, sometimes retains higher quality standards compared to fresh fish (Faruque et al., 2012). Fish drying is an age old practice followed for preservation of fish. Over the years, it has grown from a subsistence occupation to a full flourishing business (Sanjeev et al., 2020). This traditional method is followed for the preservation of fish especially in rural areas (Chakrabarti and Varma, 1999). Drying method is considered as the least expensive method of fish preservation (Balachandran, 2001). Dried fish attracts greater demand during fishing ban period or lean seasons, as the fresh fish availability is relatively low (Das et al., 2013). Dry fishes are used as food and the deteriorated Dried fish business can play an important role in the national economy in multiple ways including minimizing post-harvest losses of fish providing a source of animal protein, enlarging livelihoods, utilizing idling labor into the production process and as a dish, especially for rural poor (Wickrama et al., 2021). Dry fish marketing chain begins with a farmer (dry fish producer) and concludes with the final consumer with several intermediaries in between. These marketing middlemen provide services for producing, assembling, loading, packaging and transporting dry fish, resulting in cost increase at every stage of marketing processing (Mallick et al., 2022). Dried fish are usually low in price, easily divisible and can be stored and transported for accessibility by the poorest and most vulnerable communities throughout the year. Dried fish can therefore play a vital role in nutrition security. In India, dry fishes are widely sold at local markets and commercially important species are also exported to other countries (Immaculate et al., 2013). Dry fish are used for production of poultry/fish feed. Today, dried fish products business provide employment for millions of people, particularly women, who are the majority of the fish- drying workforce in many locations (Belton et al., 2022).

The drying techniques used for various fish employs variety of modifications in fishes, such as dehydration, weight loss, nutrient concentration, and the prevention of pathogenic microbe growth. Dried fish products are widely available, easily accessible and of great importance to the nutrition of the most vulnerable because their ease of availability intense flavour, and low prices low compared with fresh fish (Belton et al., 2022). The simplicity of storing and shipping dried fish allows it to reach remote areas where fresh fish is not easily accessible.

Dried fish is crucial for nutrition, the economy, society, and culture for many of Indian state's most vulnerable populations. In spite of this, research and policy rarely mention the broad and intricate industry that produces and distributes dried fish or the dangers it faces. A small amount of Indian fisheries catches, particularly those that are landed in coastal India, are preserved using easy techniques including sun-drying, salting, fermenting, and smoking. These processing methods can be mixed and matched based on consumer demand.

One of the main dryfish production hubs in Odisha is Paradeep. Fish that has been dried are highly sought after both inside and outside the state. Rourkela, Anugul, Sambalpur, Baragarh, and some areas of western Odisha, in addition to West Bengal, Kerala, Chhattisgarh, and Assam, have strong demand. For several decades, more than 15,000 people from the Telegu and Bengali populations have operated a dried fish business in Paradeep, and some 500 families are indirectly employed by the industry. Each year, the region produces about 5,000 tonnes of dry fish. Today, dried fish serves a variety of markets, including those for high-quality fish and prawns intended for human consumption and those for low-value fish used to make fish and poultry feed (Mallick, 2020).

The definitions of the many varieties of dried fish products are somewhat vague and inconsistent, which is understandable given that these products are frequently produced using multiple, overlapping, and non-standard processes. For instance, dried fish is frequently salted before drying, and some items that are typically regarded as salted may also have undergone a little amount of fermentation. All items can be grouped as either dried or fermented at the most general level of product category categorization. Fish sauce is a byproduct of the fermentation of fish, whereas smoked fish is dried using smoke rather than sunlight. However, product types are divided into six groups (Dried fish, Salted fish, Fermented fish, Smoked fish, Fish sauce and others) that are fine-grained enough to reflect significant regional differences in production methods.

Some of the value added dry fish such as *hentak*, an Indian fermented sun-dried powdered fish (Aarti et al., 2017); *suan yu*, a Chinese fish snack prepared by combining fermented then dried fish pieces with spices and maize meal (Zeng et al., 2016); and *ngachin*, a small fish fermented with cooked rice in Myanmar (Moe et al., 2015) are delicacy and liked by many food lover.

Table 4.1: Various dry fish available in market dry fishes

Sl.No.	Local Name	Scientific Name	Used as
1	Loita	*Harpadon nehereus*	Chutney, Curry
2	Kechhki	*Corica soborna*	Curry
3	Eaccha/ Chingri	*Leander serratus*	Fry, Chutney
4	Tangra	*Mystus cavasius*	Curry
5	Chhuri	*Lepturacanthus savala*	Curry
6	Phaluya	*Notopterus notopterus*	Chutney
7	Chandana	*Tenulosa toil*	Curry
8	Basapati	*Devario devario*	Chutney
9.	Chauli /bali kokoli	*Stolephorus indicus*	Fry
10	Ollei	*Coilia dussumieri*	Chutney
11	Mahurali	*Amblypharyngodon mola*	

Sl.No.	Local Name	Scientific Name	Used as
12	Luni kokoli / kharpania	*Ilisha elongate*	Curry
13	Nadiakhai	*Gerres filamentosus*	
14	Karandi	*Puntius chola*	Fry
15	Kharpani	*Dussumieria acuta*	
16	Marua	*Rastrelliger Kanagurta*	Curry
17	Alana kokoli	*Clupea pallasii*	Fry
18	Khanaga	*Mugil cephalus*	Curry
19	Pamplet	*Parastromateus niger*	
20	Gaanta / Minjiram	*Scomberoides lysan*	
21	Kantia / sunga	*Arius thalassinus*	Curry
22	Chitra gulura	*Nibea maculata*	
23	Pohala	*Cirrhinus reba*	Fry
24	Baunsa patri	*Salmostoma bacaila*	
25	Kumuti muna	*Loligo duvauceli*	
26	Kani	*Thunninis*	
27	Para	*Scomberoides commersonnianus*	Curry
28	Pola gara	*Gudusia chapra*	Curry
29	Gangei todi	*Xenentodon cancila*	Curry
30	Chara	*Engraulis ringens*	Curry

Drying on ground

Rack Drying

Scaffold drying

Solar fish dryer

Solar-LPG Hybrid Dryer (designed by ICAR-CIFT)

Packaged Dried fish products

Fig. 4.1 Various procedures involved in fish drying and processing

Table 4.2: Preparation of Sidal Chutney

Ingredients	Amount	Procedure
Sidal	25 g	Sidal is initially cleaned with tap water and set aside. There are tiny chunks of tomato and onion. The two are combined to make a paste. Oil is poured into a deep fryer, followed by the addition of onion chunks and a timer to let them fry. As you keep stirring, add the garlic and chilli paste to the mixture above. Continually stirring is then used to add salt, Sidal, and turmeric powder to it. Once the tomato has been added, the mixture continues to sauté until it has thoroughly boiled and thickened into gravy. Prepare it for consumption by tasting some salt.
Tomato (Raw)	150 g	
Mustard Oil	125 ml	
Onion (Raw)	150 g	
Garlic (Raw)	50 g	
Chili (Raw)	25 g	
Turmeric powder	1 tea spoonful	
Salt	A/c to taste	

Table 4.3: Preparation of Sidal Curry

Ingredients	Amount	Procedure
Sidal	50 g	Sidal is first rinsed with tap water and set aside. Cut into little pieces are the pumpkin vines, brinjal, potatoes, and onions. It is made into a paste with garlic and chile. Onion chunks are placed in heated oil in a deep fryer pan, and then it is left to fry. With constant stirring, the garlic and chilli paste are added to the mixture. Then, while continuously stirring, turmeric powder, salt, and sidal are added to it. Then, as it is being fried, chunks of pumpkin vine, potato, and brinjal are added to it while stirring continuously. After a thorough fry, enough water was added, and the mixture was then brought to a boil until all the ingredients were well cooked. Take a salty taste and get it ready to consume.
Pumpkin Vine	250 gm	
Brinjal	250 g	
Potato	250 g	
Mustard Oil	150 ml	
Onion	150 g	
Garlic	50 g	
Chili	25 g	
Turmeric powder	1 tea spoon full	
Salt	According to taste	
Water	According to need	

Entrepreneurship opportunities in dry fish sector

India is a country with unique market dynamics and high entrepreneurial spirit. During recent years, the country's industry sector is dominated by innovation led business enterprises that are found to achieve high profits within a short span of time (Arakkal et al., 2020). Fish drying over the years, has grown from a subsistence kind of occupation to a full-fledged flourishing business (Das et al., 2013). Dried fish was the primary form in which fisheries catches were exchanged and consumed before to the development of ice-making technology and cold chains. The global historical record is replete with evidence of extensive dried fish production, storage, trade, and eating dating back millennia (Fagan, 2017). However, dried fish is not a residual category. It continues to be a staple diets and cuisines throughout most of the world, and it is one of the primary ways that fish is marketed and consumed in places like South East Asia and (Liverpool-Tasie et al., 2021). Sun drying of fish is though most ancient method but it takes long time therefore method of salt curing is employed in many parts of India (Govindan, 1985) to remove moisture from fish body, therefore fishes are either brined or wrapped in salt powder before sun drying (Sugphapalaet al., 2012).

Marketing costs associated with dried fish include expenses such as transport, levy and market fees, purchasing of polypropylene bags, ice, electricity, hired labor, storage etc. The costs of fish marketing depend on the marketing channels of dry fish production

There are three observed channels which are eminent in many costal and north eastern states:

a) Dry fish Producer → Consumer

b) Dry fish producer → Retailer → Consumer

c) Dry fish Producer → Wholesaler → Retailer → Consumer

Drying is one of the most affordable post-harvest methods to preserve fish. Especially drying low value species extends the scope to avoid the fishes being spared for production of fishmeal. Dry fish production and trade offer a sustainable employment and income to the coastal fisherwomen. The production of dried fish provides jobs, a means of income, and support for millions of people. There are both large- and small-scale fishers, processors, and dealers involved from massive urban wholesalers to modest urban and rural retailers. Women are typically in charge of hot smoking and drying operations, which are typical of small-scale fisheries processing in India. However, the economy of dried fish has numerous sustainability issues, to which it also contributes. They consist of the following:

First, many workers in the value chains for dried fish come from marginalized groups (such as lower castes, widows, refugees, and members of certain religious groups), making them susceptible to various forms of exploitation as well as risks to their health and safety (Belton et al., 2019).

Second, the widespread use of pesticides during the drying and storage of fish to prevent bug infestation is believed to offer substantial health hazards to both producers and customers who are regularly exposed to these toxins. At production locations, sanitary conditions are frequently deplorable.

Third, some fisheries that are used to produce dried fish are at risk of losing their ecological integrity, which cause fish shortages and price spikes.

Fourth, using marine small fish to make feeds for expanding aquaculture and livestock industries could prevent dried fish from entering human food chains (Isaacs, 2016).

Fifth, the dried fish industry is often highly marginalized and out of the reach of formal governance structures and policies. It also lacks political representation, making it susceptible to development processes that could endanger its long-term survival (Salagrama & Dasu, 2021).

Conclusion

There immense potential of growth and development in the dry fish sector business. However, ensuring the quality and hygiene are important determinant which can limit its wider acceptance.

References

Aarti, C., Khusro, A., Varghese, R., Arasu, M. V., Agastian, P., Al-Dhabi, N. A., ... & Choi, K. C. (2017). In vitro studies on probiotic and antioxidant properties of Lactobacillus brevis strain LAP2 isolated from Hentak, a fermented fish product of North-East India. LWT, 86, 438-446.

Arakkal, R. M., Samuel, M. P., Ninan, G., & Nagarajarao, R. C. (2020). Institutional Innovations for Management and Commercialisation of Fishery Technologies in India-A Case Study. Journal of Research Management and Governance, 3(1), 18-35.

Balachandran KK (2001). Post-harvest technology of fish and products, Daya Publishing House, Delhi-110035, p 77.

Belton, B., Johnson, D. S., Thrift, E., Olsen, J., Hossain, M. A. R., & Thilsted, S. H. (2022). Dried fish at the intersection of food science, economy, and culture: A global survey. Fish and Fisheries, 23(4), 941-962.

Belton, B., Johnson, D. S., Thrift, E., Olsen, J., Hossain, M. A. R., & Thilsted, S. H. (2022). Dried fish at the intersection of food science, economy, and culture: A global survey. Fish and Fisheries, 23(4), 941-962.

Belton, B., Marschke, M., & Vandergeest, P. (2019). Fisheries development, labour and working conditions on Myanmar's marine resource frontier. Journal of Rural Studies, 69, 204-213.

Chakrabarti R, Varma PRG (1999). Halotolerant fungi in salted and dried fish at lower Visahapattinam coast.Fish. Tech. 36:28-31.

Das, M., Rohit, P., Maheswarudu, G., Dash, B., & Ramana, P. V. (2013). Overview of dry fish landings and trade at Visakhapatnam Fishing Harbour. Marine Fisheries Information Service; Technical and Extension Series, (215), 3-7.

Das, M., Rohit, P., Maheswarudu, G., Dash, B., & Ramana, P. V. (2013). Overview of dry fish landings and trade at Visakhapatnam Fishing Harbour. Marine Fisheries Information Service; Technical and Extension Series, (215), 3-7.

Fagan, B. M. (2017). Fishing: how the sea fed civilization. Yale University Press.

Faruque, M. O., Nazrul, K. S., Tonny, U. S., Islam, K. R., Dey, S. C., Mona, S. J., & Saha, D. (2012). Status of an ideal dry fish market of Bangladesh: A study on Asadganj dry fish market, Chittagong. International Journal of Life Sciences Biotechnology and Pharma Research, 1(3), 214-225.

Govindan TK, 1985. Fish Processing Technology. Oxford and IBH co. Pvt. Ltd., Tanpath, New Delhi, 137-143

Isaacs, M. (2016). The humble sardine (small pelagics): fish as food or fodder. Agriculture & Food Security, 5(1), 1-14.

Liverpool-Tasie, L. S. O., Sanou, A., Reardon, T., & Belton, B. (2021). Demand for imported versus domestic fish in Nigeria. Journal of Agricultural Economics, 72(3), 782-804.

Mallick, P (2020). Diversity of dry fish (sukhua) available in odisha and evaluation of their nutritional potential. Ph. D. Thesis.

Mallick, S., Gogoi, R., Deka, R. S., Deka, P. M., & Hoque, J. (2022). Study on North-East India's Largest Dry Fish Market, its Production and Marketing Channels. Scientists Joined As Life Member Of Society of Krishi Vigyan, 301.

Moe, N. K. T., Thwe, S. M., Shirai, T., Terahara, T., Imada, C., & Kobayashi, T. (2015). Characterization of lactic acid bacteria distributed in small fish fermented with boiled rice in Myanmar. Fisheries Science, 81(2), 373-381.

Sajeev, M. V., Mohanty, A. K., Sajesh, V. K., & Rejula, K. (2020). Dried fish consumption patterns in selected districts of Kerala.

Salagrama, V., & Dasu, A. (2021). Living on the Edge: Perspectives of the small-scale women fish processors of northern coastal Andhra Pradesh, India (Working Paper No. 07; Dried Fish Matters, p. 81). The University of Manitoba/District Fishermen Youth Welfare Association.

Sugaphapala RMNS, Suntharabathy TV, Edirisinghe U, 2012.Salt based dry fish processing and marketing by fishers of Minneriya reservoir in Sri Lanka, Trop. Agri. Research, 23(4):357-362.

Wickrama, P. S. S. L., Koralagama, D. N., & Sandika, A. L. (2021). Assesing seasonal price behaviour of selected dried fish varieties in Sri Lanka.

Zeng, X., Chen, X., & Zhang, W. (2016). Characterization of the Microbial Flora from S uan Yu, a C hinese Traditional Low-Salt Fermented Fish. Journal of Food Processing and Preservation, 40(5), 1093-1103.

5

Entrepreneurship Opportunities in Cage Farming in India

M.A. Hassan, Gunjan Karnatak and B.K. Das

ICAR-Central Inland Fisheries Research Institute Barrackpore, Kolkata-700120 West Bengal

Abstract

Fisheries is one of the fastest growing food sectors in India providing nutritional and livelihood security to millions of rural populaces. The inland fisheries and aquaculture contribute about 75% of the total fish production in the country. The inland open water fisheries resources especially reservoirs have immense scope for enhancing fish production in the country through scientific interventions and management practices. The technique of intensive fish farming through cage culture has potential to revolutionise fish production from these underutilized open water resources. This easy to adopt fish farming technology implemented by Government of India and line department of various state Governments in the form of various innovative schemes for fisheries development with initial monitory support has been rapidly adopted by farmers over the last decade. The rapid expansion of cage culture in the country will not only enhance fish production from reservoirs but also bring ample livelihood opportunities for landless fishers/farmers. The development of cage aquaculture sector beside bringing farming opportunities in open water resources will also open avenues for entrepreneurship development in allied sectors like seed, feed, infrastructure, automation, marketing etc. The cage farming technology offers equal opportunities for a larger section of educated and trained youth. The women entrepreneurs will find opportunities especially in farming and processing sector in cage fish farming. Although it is a sunrise fisheries sector, it faces several challenges and issues that need to be addressed for sustainable development. The present communication discusses present status, scope, challenges, and entrepreneurship opportunities in cage farming and way forward for development of this sector.

Keywords: Cage farming, Cage culture, Entrepreneurship, Reservoir fisheries

Introduction

Fisheries is a fast-growing and important sector for providing nutritional security and supporting livelihood in many countries. The global fish production currently stands at 179.9mt with a contribution of 112mt from marine and 65.43mt from inland sector (FAO, 2022). India is no exception to this; fisheries provide nutritious food and livelihood to millions. Our country ranks second in the world next to China in total fish production and aquaculture production. It contributes 14.7% to world fish production. Currently, fisheries is one of the fastest growing food production sectors in India with annual growth rate of 10.34% and annual production of 16.25mt. It contributes 1.1% to the national GDP, 6.7% to the agricultural GDP and provide livelihood to about 14 million. The inland fisheries and aquaculture contribute 12.12mt to the total fish production (DAHDF, 2023).

Fish production in India has gradually and definitively shifted from marine to inland fish production due to the rapid growth of the inland fisheries sector. Aquaculture development, including refinement of production protocols, introduction of new species and culture system, concomitant growth in input supply (seed, feed, medicine and chemicals), emphasis on education, training and skill development, intensive network of extension system and policy support has led to a notable increase of inland aquaculture, which currently contributes to more than 70% of inland fish production of the country. In addition, high profits in aquaculture coupled with limited water and land resources have led to system intensification using high rearing density and high protein feeds to boost productivity in closed or semi-closed systems (Zaki et al., 2020). In order to reduce mounting pressure on land resources, vis-a-vis to meet the country's fish demand, the production must come from comparatively under-exploited fisheries resources through technological interventions. India has abundant and diverse inland aquatic resources with huge fisheries potential (Sharma et al., 2010). Reservoirs are the most important and potential inland water resources (3.5mha) that can play a major role in boosting fish production in the country. However, the recent scrutiny shows the average fish yield from this mega resource is 110kgha^{-1} (Sarkar et al., 2018).

Fisheries development was or has never been the primary objective of reservoir creation in India. Therefore, there lies an inherent impediment in fisheries development in this multi- stake, multipurpose open water natural resources. The unprotected hydro-electric input channel, sluice gate or spillway is the unprotected open passage for free escape of fishes. The tall uncut tree trunks, the bamboo bushes of pre-impoundment forests pose a serious problem for gear operation leading to poor fish harvest. The un-organised displaced farmers turned fishers lacking fisheries skill, economically weak to adopt and implement technological implementation have turned the resource a subsistence income source. The harvest was therefore very meagre with their limited skill, and application of primitive craft and gear. In addition, absence of modern fishing crafts mounted with sonar system to identify fishing ground or fish finder, no defined landing centre and poor marketing channel, lack of post harvest preservation, processing or value addition are some of the glaring issues that need to be addressed.

In order to address these inherent problems of low fish yield from the vast reservoirs or similsr open water resources the academic institutions, ICAR institutues particularly CIFRI during its several decades of research endeavours, field trials and demonstrations have developed several technologies, package of practices, management protocols, for increasing fish yield. Formation of policy guidelines, fishers' cooperatives, proactive line department, fund allocation and involvement of entrepreneurs, adoption of technologies and management practices specially culture-based capture fisheries have led to yield improvement in many reservoirs. Several advantages of recent technological development of intensive fish production system known as cage culture for reservoir ecosystem have drawn the attention of policy makers, funding agencies and entrepreneurs. Considering this massive resource size and their economic and livelihood opportunity potential, the policy makers have allocated substantial amount of fund under an ambitious project, Pradhan Mantri Matsya Samridhi Yojna (PMMSY), for comprehensive and strategic development of inland fisheries involving potential states and entrepreneurs. The plan offers opportunities for entrepreneurs to initiate and undertake innovative programmes to enhance fish production from these potential resources through the implementation of scientific management plans and technological interventions. In this context, enclosure culture technology especially cage culture and its related value chain such as seed, feed, chemicals, medicine, post harvest preservation and processing have good scope of improving fish yield and entrepreneurship opportunities from reservoirs and wetlands resources of the country.

Cage farming

Cage culture is a fish production system where fish are reared in high densities in an enclosure by maintaining water exchange (Karnatak et al., 2020). Cage culture in underutilized fisheries resources can help to close the gap between the demand and supply for high-quality low fat animal protein and provide a means of subsistence to fishers in developing nations (Hassan et al., 2017).

Although relatively new in the country, it opens new opportunities for increasing fish production from vast reservoirs in the country (Karnatak et al., 2014). In addition to supplying high-quality protein, cage culture helps fishers acquire new skills and encourage entrepreneurship, increasing their income. Scientific implementation of cage culture in Indian reservoirs can contribute to another blue revolution in the inland fisheries sector (Sarkar et al., 2018).

The cage culture technology involves holding of fishes in an enclosure and managing them till they attain marketable size or age. The recent concept of cage fish farming in battery of cages in India initially started with bamboo frame now have several options for cage fabrication (Manna and Hassan 2004). The cage frame can be made up of GI pipes, HDPE blocks, PVC frames etc. The whole frame rest on floats made up of empty PVC drums or blocks to form a rigid structure (GI frame) or flexible pontoon (PVC frame or HDPE blocks). For production of table fish generally fingerling/advanced fingerlings are stocked at the onset of summer, while fry are stocked during summer/ monsoon for fingerling/advanced fingerling production. The stocking density varies with species, size or age of fishes. In case of table fish production of *Pangasianodon* sp.the recommended stocking density ranges between 30-40/ m^3 depending on water productivity. The growth of fishes is solely depended upon supplied artificial diets mainly floating feeds for different life stages fed at different times of a day (8 am & 3 pm) at variable ration levels (7-3% of fish biomass). Regular water quality monitoring, prophylactic measures and regular health check up and intermittent treatments will help disease free good crop of fishes.

Status of cage culture in India

The development of freshwater cage culture in India was initiated in 1970s in swamps using air-breathing fishes, and later in rivers and reservoirs using carps, tilapia and snakeheads (Das and Das, 2021). In India, from the initial experimental concept of individual cages to the modern concept of cage farming in a cluster or battery of floating cages was successfully designed, implemented and standardized by ICAR-CIFRI using bamboo made frame, drum float and HDPE net cage in a deep wetland of Assam for seed raising of Indian major carp, minor carp and Chinese carp (Manna and Hassan 2004). Other ICAR institutes such as ICAR-CIFE and ICAR-CIFA also experimented to raise fingerlings of carps, catfish and mahseer, later on, the same design of bamboo made floating battery of cages was implemented in Pahuj (Uttar Pradesh) and Dahod Reservoir (Madhya Pradesh) for successful production of carp fingerlings (Bhaumik and Das, 2011).

Pontoon cage

Circular cage

Fig. 5.1 Different types of cages

At present, enclosure fish farming for seed raising and table fish production has drawn the interest of developmental organizations, industrialists, policymakers, and researchers across the nation. Freshwater cage culture in open waters gained further impetus with the launching of National Mission on Protein Supplement (NMPS) in 2010 with financial aid from National Fisheries Development Board (NFDB), Government of India, through various schemes (Pandit et al., 2021). ICAR-CIFRI started developing protocols for commercial grow out cage culture of economically important species using galvanized iron cages at Maithon Reservoir, Jharkhand from the year 2011. This humble begining has led to widespread adoption of cage culture technology in more than 20 States of our country. Cage culture has been widely adopted in Odisha, Jharkhand, Chhattisgarh, Madhya Pradesh, and Maharashtra, and upscaling is currently taking place in these states (Das and Sharma, 2015). The NFDB has formulated guidelines for cage culture in inland open waters of India in collaboration with ICAR-CIFRI and Rajiv Gandhi Centre for Aquaculture (RGCA), Tamil Nadu, for the sustainable development of cage culture in the country (DAHDF, 2016). At present around 25000 cages are in operation in reservoirs across the country. Eastern Indian states mainly Jharkhand and Odisha, contribute around 43% of the total cages. The western region comprising, mainly Gujarat, Maharashtra and Goa, constitutes approximately 33% while, the North-eastern, Central, Southern and Northern states constitute 9.48%, 8.51%, 5.12% and 0.75% respectively. Galvanized iron cages form 30% of the total cages in the country.At present, Pangasius is the most dominant species, followed by Tilapia. The B:C ratio of about 1.5 have been achieved by progressive states like Jharkhand and Chhattisgarh where Pangasius is fetching a better price. It is estimated that, at an average production rate of 3 tons per cage (96m^3), the cage culture is producing about 45000 tons of fish from the reservoirs and providing employment @ of 7.4 lakh man days per year (Pandit et al., 2021).

Several fish species like *P. hypophthalmus*, GIFT Tilapia (Monosex), *Puntius javanicus, Puntius sarana,* Jayanti Rohu (genetically improved *Labeo rohita*), *L. bata, Lates calcarifer, Macrobrachium rosenbergii,* besides air-breathing fishes, ornamental fishes, murrels have also been tried in cages; however, only *P. hypophthalmus* and Tilapia (GIFT, monosex) is being produced at commercial level (Karnatak et al., 2021). The sudden and rapid expansion and high production of this species (pond and cage) has led to price crashes affecting profitability (Krishna et al., 2011, Karnatak et al., 2020). Looking at the urgent need for diversification, attempts have been made to develop culture protocols of high value indigenous species like *Ompok bimaculatus* (Karnatak et al., 2021), *Labeo bata* (Yengkokpamet al., 2020, Karnatak et al., 2020, Debnath et al., 2022), *Puntius sarana* (Upadhyay et al., 2022), *Labeo rohita* (Swain et al., 2022), *Osteobrama belangari, Labeo rohita* (Jayanti), *Heteropneustes fossilis* etc. (ICAR-CIFRI, 2021)

The state fisheries departments play a major role in the implementation of cage culture in the country with technical support from ICAR-CIFRI, Barrackpore. Cage technology has been shown to increase family income and reduce occupation migration from 29 to 9% in Jharkhand (Pandit et al., 2019)

Prospects of cage culture development

India has 37.37 lakh ha area under reservoirs of which >20 lakh ha is under medium (1000-5000 ha) and large (>5000 ha) category reservoir. Around 13982 cage units were installed in reservoirs of 23 states across country under blue revolution scheme from 2015 to 2020 (DAHDF, 2022). Another estimate reveals that there are more around 20,000 cages in Indian reservoirs including some deeper wetlands (Das et al., 2021). Recently, provision has been made for another 20,000 cages under PMMSY leading to increase in number of cages to a toal of more than 25,000. Looking towards sustainable development and ecological integrity and understanding the aspects of eutrophication due to intensive farming, ICAR-CIFRI and NFDB, Hyderabad has recommended only 0.1% of the total area under large water bodies for consideration for cage culture activity (DAHDF, 2016). Accordingly, little above 2,000 ha will be available to be covered under cage culture (Table 1). There will be 400 cages to be accommodated in 1 ha area at size groups of 6x4 or 5x5 m^2 and thus, a total of 8.6 lakhs cages could be installed in Indian reservoirs.

Table 5.1: Potential of cage culture in medium and large reservoirs

State/UT	Total area (medium and large) in ha	0.1% area (ha)	Total number of cages	Seed requirement (million nos.)	Feed requirement (tonnes) @ FCR 1.5	Potential production (tonnes)
Andhra Pradesh	160907	161	64363	257	289632.6	193088
Telangana	168000	168	67200	269	302400	201600
Arunachal Pradesh	0	0	0	0	0	0
Assam	2000	2	800	3	3600	2400
Meghalaya	1330	1	532	2	2394	1596
Manipur	0	0	0	0	0	0
Mizoram	7800	8	3120	12	14040	9360
Nagaland	0	0	0	0	0	0
Tripura	0	0	0	0	0	0
Sikkim	0	0	0	0	0	0
Bihar	5840	6	2336	9	10512	7008
Jharkhand	85911	86	34364	137	154639.8	103093
Goa	0	0	0	0	0	0
Gujarat	202106	202	80842	323	363790.8	242527
Haryana	0	0	0	0	0	0
Himachal Pradesh	41364	41	16546	66	74455.2	49637
Jammu & Kashmir	0	0	0	0	0	0
Karnataka	208634	209	83454	334	375541.2	250361
Kerala	26724	27	10690	43	48103.2	32069
Madhya Pradesh	449761	450	179904	720	809569.8	539713
Chhattisgarh	30501	31	12200	49	54901.8	36601
Maharashtra	154235	154	61694	247	277623	185082
Odisha	132151	132	52860	211	237871.8	158581
Punjab	0	0	0	0	0	0
Rajasthan	99213	99	39685	159	178583.4	119056
Tamil Nadu	42799	43	17120	68	77038.2	51359
Uttarakhand	25661	26	10264	41	46189.8	30793
Uttar Pradesh	297381	297	118952	476	535285.8	356857
West Bengal	15000 (7000 suitable)	7	3000	12	13500	9000
A & N Islands	0	0	0	0	0	0
Puducherry	0	0	0	0	0	0
Total	2157318	2157	**859927**	3452	3883172	**2588782**

Source: Data updated based on Das et al., 2021

Entrepreneurship opportunities

Cage culture is a potential technology for employment generation not only for sustaining livelihoods but also for developing entrepreneurship and expanding in a big way in direct farming or allied activities. The success of cage farming depends on multiple factors such as quality seed, cost effective and quality feed, fish health management, skill/knowledge/capacity of the farmer/entrepreneur, culture protocols for diversified species.These multiple factors have the potential to develop into allied sectors, such as, live seed producer, trade and transport, feed raw materials, feed manufacture, trade and transport, live fish transport, post harvest preservation, processing, value addition, product diversification and trade can contribute to the success of cage culture and offers scope for entrepreneurship development. Entrepreneurship is the process of identifying and utilizing available opportunities and resources to convert an idea into the form of a product or service to market (Sukham, 2022). Therefore, in cage culture sector an individual can find opportunities in both farming or in any of the allied sectors of value chain.

Good quality seed is a prerequisite for successful cage farming. Timely availability of good quality seed is a big challenge in cage farming especially when it comes to seed availability of high value species. The country produces 540960 lakh fry per annum, most of which is carps. The demand for seed is so high that, a part of the seed requirement of high value fish species like *Clarias magur, Heteropneustes fossilis, Ompok bimaculatus* and even bulk demand for Pangasius seed are met from unauthorised sourced seed from across the Indian border. Indigenous seed production of these high value species or bulk demand of Pangasius to meet the seed requirement for cage farming in reservoirs will open a new dimension of entrepreneurship. Cage culture requires supply and transport of advanced fingerlings across the country from the seed hub in West Bengal for table fish production. Venturing into hatchery sector or advanced fingerling production in different regions of the country for creating seed hubs/cluster near cage culture sites will open another opportunity for entrepreneurship. Live seed transportation is another trade opportunity in the sector. The seed production business however requires skill, knowledge and training on brood stock management, induced breeding, hatchery management and pond management including water quality, feeding and health management.

Feed is the next most important input (55-65% of operational cost) when we talk about cage farming which is mostly feed based. Reportedly, about 30 high-tech feed mills with capacity of 2mmt per year are operational in India for fish feed production, providing employment opportunities for skilled /

non-skilled workers. A report indicates, at present feed mill utilization is only 50 % of the capacity. Looking at increasing feed-based farming, another 30 feed mills (6-10 ton/hr capacity) were sanctioned by DAHDF, covering 11 states under the Blue revolution Scheme. In addition, many low capacity feed mills (250-500kg/hr) have been and are being established under PMMSY. This presents an opportunity for entrepreneurs to venture into raw material supply, feed production and trade contributing to development of the land-based aquaculture and cage culture sector providing employment opportunities, thereby strengthening and boosting the rural economy.

Health management is another important aspect of fish farming. It is not advisable to use chemicals and drugs in open water bodies as the water from many reservoirs is used for urban supply and drinking purpose. It is especially challenging to manage health of the fishes in high density cage farming. The challenge brings with it opportunities for development of nutraceuticals or designer feeds suitable for open water cage farming with an aim to have low polluting and medicated feeds for boosting immune system of fish. Disease in fish farms causes huge monetary losses. The country still lacks fish disease diagnostic centres where farmers can test the samples and get proper advice on prevention or treatments. The setting up of such centres will not only open business opportunities for fisheries health professionals but also prevent disease outbreaks keeping the stocks healthy thereby increasing production.

As per the potential, the suitable area in reservoirs can accommodate approximately 859927 numbers of cages. At present there are only a few companies that supply the ready to install cages. There is a huge market opportunity for fabricated cages (GI/HDPE etc) in future. It also opens opportunities for venturing into business of cage material supply like PVC pipes, nets, HDPE blocks, Ropes, floats, antifouling paints etc. In this era of smart farming, development, and marketing of ICT tools like drones, underwater cameras, automated sensors for farm monitoring and management, automated feeders etc will help in revolutionising cage culture.

Most of the reservoirs are located in remote areas where marketing of fish is an issue. There is scope for marketing channels. A common platform formed by group of entrepreneurs from both production and marketing sector can make marketing of the produced fish easy by creating an interface between buyers and sellers. Creation of e- marketing portals showcasing price, size, species, availability, and delivery option at required sites will help the producers to sale their fish at competitive rates. With such facilities producers can sell and buy seed/table fish/feed/medicines/chemicals etc on pan-India basis using simple digital apps.

Women empowerment through cage culture

Women contribute equally to fisheries sector of the country. Out of the total fishermen population in the country 44% are women. Cage culture is an excellent opportunity for developing women entrepreneurs in fisheries (Joseph and Gopalakrishnan, 2017). With multiple schemes and subsidies provided by the government for cage farming many entrepreneurs and farmers have come up to make it big in fisheries sector. In recent years in states like Odisha, Jharkhand and Chhattisgarh women (individuals and SHGs) have also shown interest and encashing these opportunities and ventured into cage farming. The cage fish culture is easier to monitor and manage and do not require all day involvement for women. It can be a source of additional income for the women otherwise engaged in full time household activities. It brings equal opportunities for educated and trained youth as seen in case of cage farming in Hirakud reservoirs where cage farming has been taken up not only by fishers/ farmers but also MBA graduates and engineers under the state reservoir fishery policy. It is a potential technology for women empowerment.

Challenges/Issues

The cage culture at present is mostly funded and supported by state departments under various schemes. Despite high production potential (30-40 kg/m^3) and tolerant to high stocking density 30-40 nos./m^3) of commonly cultured species, Pangasius, in cages and availability of many schemes to promote cage culture, fish farmers and entrepreneurs are reluctant to adopt the cage culture independently. The major constraints affecting the adoption of cage culture by farmers are presented here (Figure.1.) High initial cost, high volume and cost of feed, vulnerability to disease at low temperature in early stage, comparatively low market price, lack of processing and value addition of Pangasius flesh are found to be the major constraints hampering wide-scale adoption of cage culture in the country (Pandit et al., 2021).

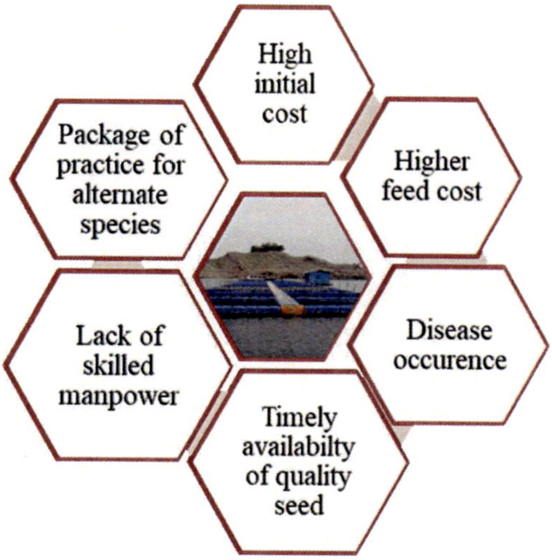

Fig.. 5.2: Major constraints and issues in cage farming

Way forward

Any entrepreneurs operate in a complex and dynamic environment. They are part of a larger collection of people including other farmers, suppliers, traders, transporters, processors and many others. Each one of them has a role to play in producing the products and moving them through the market/value chain. Each one needs to be an entrepreneur. They also need to respect each other and work together to make the whole system work better and be more profitable (FAO 2012). Considering the national target of 26mmt of fish production, there exists huge scope for entrepreneurship development through cage farming in inland reservoirs. Only large and medium reservoir can accommodate approximately 8.6 lakh number of cages. However, many State Governments have allowed cage farming in small reservoirs having sufficient depth and annual flushing of reservoir water. Compared to present number of close to 25000 cages, there lies wide gap between scope of cage operation and existing practices. Huge mandays will be needed to produce, trade, transport and install this huge number of cages. It is estimated, once the full potential will be realised in terms of cage installation only in large and medium reservoirs that can yield additional 2.5 m tons of fish. The feed requirement to support fish production exclusively from cages will create a demand for 3.75 m tons of fish feed. In order to manage and support this massive fish production system technically and financially sound entrepreneurs with risk taking and innovative capability and visionary outlook will be required. To promote cage farming, researchers have to develop environment friendly and economically viable enclosure fish culture technologies. India being a continent, region specific packages of practices, suitable strains of desired species for cage farming, brood banks and seed production technology of regionally preferred indigenous fish species, low-polluting, cost-effective and nutritionally balanced artificial feed and diversification of fish species focussing regional preference and market demand need to be developed. The challenges to reduce losses due to disease occurrence, minimise use of antibiotics, management of fish health in cages by application of nutraceuticals has to be mitigated. Proper planning, policy guidelines and better management practices are to be formulated and implemented for sustainable cage culture production and maintenance of environmental integrity.

To ensure timely availability of inputs for cage farming cluster approach in PPP mode is required to be implemented. Strengthening of the value chain through development of market links will be necessary for enhancing profitability of cage culture. Given these issues are addressed; enclosure farming has the potential to revolutionise fish production from low yielding reservoirs and double farmers income.

To realize the potential of fish production from reservoirs and wetlands the entrepreneurs' using cage culture technology are to be technically competent and innovative and capable of prior planning to steer their businesses through the stages of enterprise development – from establishment and survival to rapid growth and maturity. However, there are many challenges that these farmers may face such as social taboos/barriers, economic barriers, regulations, access to finance and information, and their own managerial capacity to cope with risks and challenges and to seize opportunities.

The inland cage culture in India is still in its infancy and needs lots of hand holdings by extension workers and training institutes. Extension institutions/ workers have to support farmers by facilitating links with financial institutions and in providing information on market opportunities and facilitating links with buyers. Entrepreneurial farmers are to be supported by facilitating linkages and developing networks and partnerships among farmers and with otheractors in the value chain. Similarly, cooperative financial institutions, conducive government policies, proactive line department will motivate the enterpreneuer to venture into a vibrant and large scale cage fish farming in the country.

References

Bhowmick U and Das A K, 2011. Cage and pen culture220-245.In: Handbook of fisheries and aquaculture. (Eds) Ayyappan A, Moza U, Gopalkrishnan A, Meenakumari B, Jena JK and Pandey AK. ICAR, New Delhi

DAHDF, 2016. Guidelines for Cage Culture in Inland Open Water Bodies of India. Hyderabad, India: National Fisheries Development Board.

DAHDF, 2022. Handbook of fisheries statistics. Department of Animal Husbandry, Dairying and Fisheries, New Delhi, India.

Das A K, Meena D K, Panda D and Sharma A P, 2014. Cage farming in Indian reservoirs: Economic viability and future prospects. World Aquaculture 45(3): 1-3.

Das A K and Das B K 2021. Evolution of freshwater cage culture in India and role of ICAR-CIFRI-retrospect and prospect. Journal of Inland Fisheries societ*y* of India, 52 (2): 123-138.

Das A K and Sharma A P 2015. Cage aquaculture in inland open waters of India: retrospect and prospect. In: Book of abstracts, 5th international symposium on cage aquaculture in Asia (CAA 5)., 25-28 Nov, Kochi, India p. 55.

Das B K, Sarkar U K, Samanta S, Das A K, Hassan MA, 2021. Strategic Plan for inland open water fisheries development under PMMSY. ICAR-CIFRI, Barrackpore p.60.

Debnath D, Das BK, Yengkokpam S, Das P, Yadav A K, Sharma N, Borah S, Ray BC., Kakati, AK, Sarkar UK and Bhattacharjya BK, 2022. Evaluating growth, production and economics of a new candidate species Labeo bata in cages: A regional model for table fish production in floodplain wetlands of North East India. Aquaculture 546: 737344. https://doi.org/10.1016/j.aquaculture.2021.737344.

FAO, 2012. Entrepreneurship in farming. (Ed. David Kahan), Food and Agriculture Organization of the United Nations, Rome.

FAO, 2020. The state of world fisheries and aquaculture, Food and Agriculture Organization of the United Nations, Rome.

Hassan M A, Mishal, P, Karnatak G and Sharma A P 2017. Toward the blue revolution in India: prospects for inland open waters. World Aquaculture 48(1): 25-28.

Karnatak G, Das B K, Mishal P, Tayung T, Kumari S, Sarkar U K, Das A K, Ali Y 2020. Impact of stocking density on growth, feed utilization and survival of cage reared minor carp, Labeo bata (Hamilton, 1822) in Maithon reservoir, India. Aquaculture, 532: 736078.

Karnatak G, Das B K, Puthiyottil M, Tayung T, Kumari S, Lianthuamluaia L, Sarkar U K, Behera B K, Tiwari V K, Chadha N K and Das A K 2021. Environmental parameters and stocking density influence growth, feed utilization and economics of butter catfish, Ompok bimaculatus (Bloch, 1794) production in floating net cages in a large tropical reservoir, India. Environmental Science and Pollution Research, 28: 59720-59730.

Karnatak G and Kumar V 2014. Potential of cage aquaculture in Indian reservoirs. International Journal of Fisheries and Aquatic Studies 1(6): 108-112.

Krisnha R R, Rao K VP and Reddy D M2011. The princes of aquaculture and the plights of the fish farmers. Aquaculture Asia, 16 (2): 12-16.

Manna R K and Hassan M A 2004. Cage culture of fish – A handbook. CIFRI bulletin no.132, Central Inland Fisheries Research Institute, Barrackpore, pp.26

Pandit A, Das B K, Chandra G, Roy A, Debroy D, Yadav A K, Chakraborty L and Biswas D K 2021. Indian Journal of Fisheries 68(1): 76-81.

Pandit A, Debroy P, Chandra G, Roy A, Chakraborty L, Das B K and Biswas D K, 2019. Adoption of cage farming has enhanced livelihood security of the reservoir fishers: evidence from Jharkhand state of India. Journal of Inland Fisheries Society of India, 51(2): 184-189

Sarkar U K, Sandhya K M, Mishal P, Karnatak G, Lianthuamluaia Kumari S, Panikkar P, Palaniswamy R, Karthikeyan M, Mol S S and Paul T T, 2018. Status, prospects, threats, and the way forward for sustainable management and enhancement of the tropical Indian reservoir fisheries: an overview. Reviews in Fisheries Science and Aquaculture, 26(2): 55-175.

Sharma A P, Jha B C and Kaitha P K 2010. Draft report: Midterm evaluation of NFDB program on stocking of fingerlings in reservoirs and its impact. ICAR-Central Inland Fisheries Research Institute, Barrackpore.

Sukham M, 2022. Fish Nutrition, Feeds and Feeding-An Aquapreneur's Perspective. In: Promoting AquapreneurshipthroughPotential Aquaculture Systems (eds. Priyakumari et al). Madras Christian College, Chennai & MANAGE, Hyderabad, ISBN: 978-93-91668-25-9. p.11-18.

Swain HS, Das BK, Upadhyay A, Ramteke MH, Kumar V, Meena DK, Sarkar UK, Chadha NK and Rawat, KD 2022. Stocking density mediated stress modulates growth attributes in cage reared Labeo rohita (Hamilton) using multifarious biomarker approach. Scientific Reports, 12(1), pp.1-14.

Upadhyay A, Swain H S, Das B K, Ramteke M H, Kumar V, Krishna G, Mohanty B P, Chadha, NK and Das A K 2022. Stocking density matters in open water cage culture: Influence on growth, digestive enzymes, haemato-immuno and stress responses of Puntius sarana (Ham, 1822). Aquaculture, 547: 737445.

Zaki M A, Alabssawy A N, NourA E A M, El Basuini M F, Dawood M A, Alkahtani S and Abdel-Daim M M, 2020. The impact of stocking density and dietary carbon sources on the growth, oxidative status and stress markers of Nile tilapia (Oreochromis niloticus) reared under biofloc conditions. Aquaculture Reports, 16: 100282.

6

Entrepreneurship Opportunities in Fish Processing Sector in India

Ranendra Kumar Majumdar

Department of Fish Processing Technology & Engineering, (Former Professor and Head), CAU - College of Fisheries, Lembucherra- 799210, Agartala

Abstract

During the last decade, the fish processing industry in India is on an assured track of growth and profitability. Fish and fish products have presently emerged as the largest group in agricultural exports of India, with 17.35 lakh tonnes in terms of quantity and Rs. 63,969.14 crores in value during FY 2022-23 despite the several challenges in its major export markets like USA. Last two decades also witnessed technology developments in fish processing sector and radical changes in the life style of urbanites. These along with increase in productivity, consumers demand of safe fish, increase in shelf life, improved food safety and reduce waste during processing operations, offer abundant scope for innovation which lead to entrepreneurship. The commercialization of fish products still poses lot of challenges to the entrepreneur and researcher in terms of optimization of technologies and ultimately developing the technologies into a commercially viable business plan.

Keywords: Fish products, Fish by-products, Fish processing, Fish processing technology, Sidal

Introduction

With the growing population of the country and the increasing requirements for fish protein, the need for sustainable development of the post-harvest fisheries sector is now felt much more than ever before. Fish and fishery products serve as a part of a balanced diet because fish is known to be an excellent source of proteins, essential fatty acids, vitamin D, vitamin B12, and many useful trace elements such as selenium and iodine. Especially, Omega-3(n-3) fatty acids such as DHA and EPA are plenty in fish, contribute significantly for the functioning of vital organs viz heart, brain, immune and reproductive systems.

Over the years the demand for fish is on the rise all over the world due to such health benefits.

The harvest and post-harvest handling, processing and distribution of fish provide livelihood for millions of people as well as provide valuable foreign exchange earnings. India's scale, its large and growing population and its over 500 million strong labour force pose a daunting challenge to job creation. The rising population in India is also accompanied by the rise of unemployment in all the sectors. Indian economy is also being hampered due to this problem. Meanwhile, it is also not possible for any government to create job for all. Probably, the solution of this burning problem is to encourage and support entrepreneurship.

Entrepreneurship is a dynamic process of creating incremental wealth and impact, and entrepreneurs are people that notice opportunities and take the initiative to mobilize resources to make new goods and services. Over the years, entrepreneurship has played a vital role in shaping the Indian economy, and it is set to play an even more important role in the future. Startups and small businesses are often more nimble and able to adapt to changing market conditions, which allows them to bring new products and services to market more quickly. The government has introduced a number of programs and initiatives aimed at promoting entrepreneurship, such as the Start-Up India program and the Pradhan Mantri Mudra Yojana. These programs provide financial and logistical support to entrepreneurs, helping them to get their businesses off the ground and grow. With the government's support and a growing ecosystem of venture capital and angel investors, entrepreneurship has become an increasingly attractive career choice for many young people in India. Innovation is a key driver of entrepreneurial success, so it's important for entrepreneurs to be aware of these different sources and to find ways to leverage them in their businesses. One of the most common sources to develop innovative ideas for an entrepreneur is market and customer needs. Entrepreneurs often identify new products or services based on customer needs that are not being met by existing businesses. Entrepreneurs may learn a lot from the opinions of their customers. Entrepreneurs can find chances for innovation and create goods or services that cater to those demands by paying attention to the wants and preferences of their customers. Entrepreneurs can produce novel products, services, and processes that can provide them a competitive advantage by remaining current on the most recent advances in their field and learning new skills and techniques.

Fish processing sector of India

The Indian Blue Revolution has led to a major improvement in the fishing and aquaculture industries. The Indian fisheries industries are regarded as sunrise sectors and as being the third-largest fish producer and the second-largest aquaculture producer globally, India recognizes the significance of the fisheries and aquaculture industry. In the recent past, Indian fisheries has witnessed a paradigm shift from marine dominated fisheries to inland fisheries, with the later emerging as a major contributor of fish production from 36% in the mid-1980 to 70% in the recent past. Fish and fish products have presently emerged as the largest group in agricultural exports of India, with 17.35 lakh tonnes in terms of quantity and Rs. 63,969.14 crores in value during FY 2022-23 despite the several challenges in its major export markets like USA.

Post-harvest fisheries is defined as what culminates into activities, processes, and steps after harvesting, handling, processing, distributing, marketing, and consumption. Post-harvest fisheries is considered a critical sub-sector and plays a very significant role in contributing to national food and nutrition security, livelihood of the people, and the national economy. Post-harvest technology serves to make fishery products available at the place, at the time and in the form required. One attractive way to help fishing communities is to improve the technology they use by introducing new or better ways of processing, handling, and packing fish. As spoilage of fish starts right from the time it is caught, proper storage, preservation and prompt disposal or transport services are essential. Various studies have pointed to the high levels of wastage in the Indian fish and fishery products due to spoilage. This is particularly acute during the monsoon, when up to 15-20 percent of the catch reported to be lost.

Situation analysis

Presently, fish markets, both wholesale and retail in the country are in a pathetic condition. Besides, a larger volume of fish is sold through street markets, often on footpaths close to open drains. This unhygienic environment and the fact that fish is seldom kept in ice, results in fast deterioration of the quality of fish. Mostly whole fish is sold in the market and there is negligible processing/value addition. Further, while marketing, transportation or storage of fish, the standard norms of hygiene and sanitation are least considered, leading to a product that is contaminated and unsafe from food safety point of view. Furthermore, fish products' safety and quality control are poorly practiced in the country. Lack of cold storage facilities and fluctuating weather conditions are the major challenges impacting fish post-harvest activities. Therefore, introduction of new technology, for supplying safe fish to the consumers, preferably in their door steps are some potential areas where development of innovative ideas would help to create entrepreneurships. Significant improvements in fish trade for both local and international markets largely depend on the quality of fish.

Key issues and challenges of the fish processing sector

The key issues and challenges that are currently faced by the Indian post-harvest management of fish include: -

- Lack of adequate marketing infrastructure to hygienically handle existing quantity of fish
- Lack of infrastructure for proper fish distribution chain
- Lack of adequate cold chain facilities
- Modernization of fish marketing system
- Lack of hygienic fish landing centres
- Lack of availability of market information – price and supply
- Lack of adequate fish processing units
- Lack of entrepreneurships for value added fish products
- Production of dried, smoked and fermented fish are not organized especially in the northeastern states of India
- Thoughtful utilization of fish wastes for different valued products is not adequate
- Lack of skilled manpower to support different actors of the value chain

Judicious study of the present-day problems faced by the consumers and the fish processing sector as a whole would help to think about many innovative ideas to provide safe fish and services to the consumers, create demand of the fish products in both domestic as well as in the international markets, not only as food but also for different functional compounds.

Entrepreneurship opportunities in fish processing sector

Being a highly perishable commodity, fish require immediate processing and various options are available for the value addition of fish. Unveiling of many facts regarding nutritive values of fish in recent times has resulted peoples' consciousness regarding benefit of eating fish and thus increased its demand, either in fresh or in processed forms. Although there are many definitions, however, value added food products mean raw or pre-processed commodities whose value has been improved through any additional activity (s) such as addition of ingredients or processes or mode of consumption that make them more attractive to the buyer at the time of sale and/or provide convenience to the consumer. These products range from live fish and shellfish to ready-to-serve convenience products.

The conventional products are disappearing from the market and their place is being taken up by the hygienically prepared and attractively packed products in the form of 'ready-to-cook', 'ready-to-serve' convenience foods. The people have become more selective in their food choices and are ready to spend more on convenience foods such as assembled meals, rather than preparing from base ingredients. Such preferences are the result of continual changes in work cultures and life styles, availability of disposable income and lack of leisure for house hold work. Health oriented food and nutraceuticals are preferences of today's consumer. Followings are the drivers for entrepreneurial intervention in the fish processing sector:

- Society and lifestyle trends
- Consumer sophistication, health, knowledge
- Technological advances, industry specialisation
- Market trends, prices, wholesale, retail
- Hygiene, food safety, quality assurance
- Sustainability, environmental and ethical concerns
- Marketing, public relations and communications

For efficient resource and product diversification following factors need to be considered:

- Identification of major fish, low-cost fish, fish discards, by-catch and underutilized species
- Comprehensive study of underutilized fish and fish discards
- Optimization of technologies specific to the region for fish processing, preservation and product development
- Introduction of new products, concepts, services
- Supplying of novel and improved biological raw materials for high quality food specially from aquaculture
- Overall process optimization for upgrading of fish discards, by-catch, fish wastes and by-products

- Process optimization – advanced and optimized technologies, packaging systems and process control
- Application of the 'clean production' concept

Based on the deficiencies in the present fish processing sector of India, as well as the need of the consumers, different areas are discussed here for guiding the budding or existing entrepreneurs for intervention.

Retail sale of fish

Presently, both wholesale and retail fish markets in the country are in a pathetic condition. Besides, a larger volume of fish is sold through street markets, often on footpaths close to open drains. Further, while marketing, transportation or storage of fish, the standard norms of hygiene and sanitation are least considered, leading to a product whose quality is deteriorated fast leading to contaminated and unsafe fish from food safety point of view. Such unhygienic environment of the fish markets prevents a section of people mostly in urban and semi-urban areas from access to the fish market. Rather they prefer quality fish in their door steps. Moreover, there is a growing preference of buying fish in live condition. In case of metro cities, due to lack of spare time to purchase fish from markets, a section of people wants fish in dressed and ready-to-cook condition at any time. Therefore, from the above stated concerns and needs of the consumers, followings are the scope of entrepreneurships.

1. Door-to-door sale of fish in insulated box, where fresh fish will be kept in ice. There should be provision of dressing of fish and wastes should be collected for disposal to municipal garbage bins.
2. Door-to-door vending of live-fish using live-fish vending system with provision of dressing and proper waste disposal.
3. Fillets/steaks of quality fish (dressed, washed and weighed), packaged in sealed tray, kept at deep freezer (-20°C) will be delivered on phone call or will be available from a store in market.

Target Market / Customer

- High end customers and urban population
- Working couples

Machineries required

- Deep freezer (-20°C)
- Live-fish vending system (built on battery operated rickshaw/e-rickshaw)

- Packaging & sealing machine
- Balance

Value added fish products

As a matter of fact, whenever we speak about the scope of entrepreneurships in fish processing sector, the ideas initially come to our mind regarding value added fish products and also frozen fillets, IQF products, etc. Although, there are huge potential of entrepreneurships in the field of fish processing and only a part of it has so far been harnessed in organized way. Value is a combination of quality, service and price. The basic benefits of value addition to food include the functional and emotional benefits related to quality and nutrition, convenience in preparation and high sensory appeal at a reasonable cost. In general, value-added food products are raw or pre-processed commodities whose value has been increased through the addition of ingredients or processes that make them more attractive to the buyer and/or more readily usable by the consumer, with enhanced appeal and attraction. The dual advantages of value-added fish products are utilization of low-priced fish species and providing protein-rich convenience foods, have been considered as the major outcome of the value addition. It is a production/marketing strategy driven by customer needs and perceptions.

Usually, the fishes which are less preferred and fetch comparatively low price in fresh condition including by-catches, in case of both freshwater and marine fish/prawn species are preferentially utilized for preparation of innovative value-added products. This is possible due to modern processing facilities available for value addition to fish and such hygienically and attractively packed value-added fish products have good prospect to catch domestic as well as international markets. Value addition has opened a new horizon of entrepreneurships in the fish processing sector. Although, in India even now, fresh fish is the preference of the majority. However, there is a growing popularity of value-added fish products as evidenced by their availability in the modern super markets and malls. Although, production and marketing of value-added fish products are presently practiced for women empowerment and income generation by the SHGs, however, small or medium-scale commercial enterprises can be established based on the technology of value-added fish products and these would create the scope of employment generation in both rural and urban sector. Development of indigenous fish processing machineries is a real necessity of the hour for cost effective value addition in fisheries sector.

It is fact that bulk of the value-added fishery products exported at present come under the category of frozen products from marine fish and shrimps, and this is almost saturated due to stagnation of marine landings. However, to augment earnings from export, a most viable alternative way is the promotion of diversified value-added fish products, which will not only accelerate our forex earnings in export, but would provide a multiplier effect on employment front, especially for weaker sections and womenfolk. Diversified value-added fish products having demand in domestic market primarily fall under the following major categories.

Pickled products: Pickles from low-cost and less bony fish and shrimps have good demand in market, which claim shelf-life of six months at room temperature. Other ingredients required for fish pickle are ginger, green chili, garlic, turmeric powder, chili powder, cumin powder, coriander powder, mustard power, fenugreek powder, cinnamon and cardamon power, salt, sugar, vinegar, sodium benzoate, refined oil. Pickles are packed in sterilized glass bottle/stand-up pouches and sealed airtight.

Coated products: The most prominent among the group of value-added products is the battered and breaded products, processed out of a variety of fish and shellfish, offer a 'convenience' food, relished widely by the consumer. The production of battered and breaded products involves several stages. The method varies with the type of product and pickup desired. In most cases these steps are involved: (1) portioning/forming (2) pre-dusting (3) battering (4) breading (5) pre-frying (6) freezing, and (7) packaging and cold storage. According to requirement and demand, seasonings and spices can be added to the batter. The products, particularity the coatings are very delicate and require protection against mechanical protection to the products. Thermoformed plastic containers are most suitable for packing such products. A variety of battered and breaded products can be prepared from shrimp, squid, clams, fish fillets, minced meat from low-cost fish have good market potential.

Shrimp products: Shrimp in different forms such as peeled and deveined, butterfly, round tail-on and cooked and peeled.

Squid products: Two most demanded squid products are squid ring and stuffed squid.

Clam and other related products: Live clams and other bivalves such as oyster, mussels etc. are depurated and the meat is shucked out after boiling, and converted into coated products.

Fish mince-based products: Fish flesh separated in the form of reduced particle size from the skin, bones, scales, and fins etc. are known as mince. Mince serves as intermediate product for versatile product development. Following are some popular mince-based value-added fish products.

Fish fingers: Fish fingers are suitable for restaurant and fast-food centre and can be prepared from fish fillets or by using minced meat. Any big variety of fish preferably with less bone can be used for fish fingers.

Fish cutlets: Fish cutlets are very much popular ready to cook value added products prepared from low priced fish. It has a good demand in restaurant and fast-food centre.

Fish ball: Fish balls are steamed product made from fish mince.

Fish nuggets: Fish mince is mixed with salt and spices, steamed and can be used for making curry or battered and breaded.

Fish sandwich: Sandwiches are very delicious and popular food commodity especially in restaurant. Fish sandwich paste is made using fish mince and spices, which can also be retort processed for use as sandwich spread afterwards.

Fish momo: Presently, momos are very popular product in every fast-food stall, parks, etc. Usually, it is made from pork or chicken. However, people who are in dietary restriction to consume pork or chicken, prefer momo made from fish.

Fish wafers: It is prepared by preparing paste blending minced meat, starch, spices and salt. On frying the wafers swell to two to three times of its initial size and become crisp and delicious.

Fish papad: Fish papad is very much liked as starter and can also be consumed as a snack item before meals to increase the appetite.

Fish soup powder: Fish soup powder can be formulated from any type of fish having very low-fat content which serves as a rich source of animal protein and other nutritional factors. Cooked fish mince, starch, spices, fat, thickening agent etc. as the main ingredients.

Fish noodles: This is a product apparently similar to ordinary noodles available in the market, but enriched with protein and other nutrients of fish. The product has very good rehydration property and only two minutes cooking is required for preparation of fish noodles.

Extruded products: Fish based extruded snack-like products have got very good marketing potential. Formulation of appropriate types of products using fish mince, cereals etc., with attractive packaging can be popularized in the markets advocating its nutritional value superior to similar products available which are made only with cereals.

Besides these, many other value-added products can be developed using fish mince, such as fish samosa, fish chop, fish pakoda, fish chakli, fish bhujia, etc. Although, entrepreneur can develop many other new products through trial-and-error method and consumers' acceptance.

Surimi based products: Surimi is a Japanese term used for separated fish mince, which is water washed, partially dehydrated, mixed with additives, block frozen and frozen stored. The major advantages of surimi are that any underutilized and less-utilized fish species can be used as raw materials, and variety of value-added products can be prepared by altering the appearance and eating quality through application of various processing technologies and ingredients. In India, surimi is mostly used for production of the following popular products.

Fish sausage: The preparation of sausage primarily consists of grinding the surimi with salt (2.3 to 3%), sugar, STPP, starch, sodium glutamate, spice mixes, vegetable oil, colouring material, preservatives and condiments, blended in silent cutter and packing in casing followed by sealing and heat treatment.

Analog or imitation products: Analogue products are produced from surimi made out of under-utilized fish species, and are designed to mimic high-value, luxury food items. Shell fish analogues are popular surimi seafood products which imitate shell fishes such as crab, lobster, shrimp etc. in shape, colour, appearance, flavour and textural properties.

Frozen fillets and fish steaks: Large sized conventional fishes could he filleted, glazed, individually frozen and wrapped. These can be stored in deep freeze (-18°) for minimum three months. Large sized fishes could be sliced into thin steaks of uniform thickness using a power meat cutter. The steaks can be individually frozen and vacuum packed. Minimum shelf-life for these products is three months at -18°.

Texturized fish products: When fish mince or surimi is partially denatured using sodium chloride, form actomyosin gel, which is thermally set. The final product can be given cubical shape, packaged and stored in deep freezer at -18°C, where it is kept well for six months. The product is rich in antioxidants and other functional compounds from the natural spices, without any chemical preservatives, boneless and is having paneer type of appearance.

Ready to serve fish curry: Fish curry in ready to consume form, packed in metal cans is a popular value-added product. The sardine, seer fish, mackerel, herring etc. are usually used for this product. Because of some of the disadvantages for metal cans and flexible pouches, a three-layer configuration of flexible retortable pouches based on polyester/ aluminium foil/cast polypropylene has successfully developed. Fish curry in different culinary style can be processed in retortable pouch in over-pressure retort with different time-temperature combination. Storage life is more than one year at ambient temperature.

IQF (Individually Quick Frozen) products: In this technology, the entrepreneur would be able to supply the customer with frozen items in small ready to cook pieces instead of solid blocks, which need to be thawed prior to cooking. Other than the high initial investment cost, it also needs skilled operation and maintenance staff. This product is suitable and convenient for the growing market in semi prepared and ready to eat products. Different IQF products available in our country developed from shrimp, lobster, cuttle fish, clam, skinless fillets, etc. IQF products fetch better price than conventional block frozen materials in the domestic markets. The products are to be packed in attractive moisture proof packing and stored at -20°C or below without fluctuations in storage temperature. Thermoform trays have now become an accepted packaging container for IQF products.

Target Market / Customer

- High end customers and urban population
- Working couples
- Executives, travellers, etc.
- Strategic shopping locations such as malls, railway platform, airport, etc.

Machineries required

• Refrigerator	• Moulding machine
• Deep freezer (-20°)	• Battering and breading machine
• Meat bone separator	• Fryer, Cooker
• Mincer	• Blast freezer
• Silent cutter	• Packaging & sealing machine
• Sausage stuffer	• Air blast freezer
• Pulveriser or Grinder	• Pickle Filling and Sealing Machine
• Weighing machine	• Steamer
• Tray sealing machine	• Over-pressure retort

Marketing strategies

The main objective of product development is to generate top quality products that perform well in the market meeting the consumer needs and bring profits for the company. An understanding of market forces and elements of the marketing programme is a prerequisite for a company venturing into new products. For entry of a new product in the market, i. e. the 'First to Market' strategy is based on strong research and development and requires greater resources, long term planning, technical leadership and willingness of the company to take risk. The pattern of marketing of value-added fish products is completely different from the traditional fish trade. In case of value-added fish products, marketing is dynamic, sensitive, complex and expensive. Few important factors responsible for successful launching of any product include surveys, consumers' acceptability study, packaging, display, advertisement etc. Most of the market channels prevalent in our country are not adequate to trade value added products. An improved super market chain is required for this purpose. Moreover, advantages of modern packaging technologies must also be adopted to enhance the attraction of the products to the consumer.

Products from fish by-catches and by-products

Huge quantity of very small sized fish landed as by-catch which do not find a ready market as fresh fish. Fish processing and filleting industries turn out large quantities of fisheries waste. Fish wastes are generated at different points in the value chain, viz. by-catch, onboard handling, landing centres, transportation, storage, processing, retailers, and consumers, whereas, processing wastes are generated mainly in the fish processing plants and retail markets. Solid fish waste consists of head, tails, skin, gut, fins and frames, shells of crustaceans, and these by-products can be a great source of hi-value products, as listed below.

• Fish meal	• Fish oil	• Fish silage
• Fish manure	• Foliar spray	• Fish protein powder
• Different digestive enzymes	• Collagen	• Gelatin
• Chitin	• Chitosan	• Fish maw
• Ising glass	• Fish glue	• Pearl essence
• Shark fin rays	• Shark skin	• Shark cartilage
• Glucosamine hydrochloride	• Chondroitin sulfate	• PUFA concentrate
• Carotenoproteins	• Pigments (astaxanthin)	• Minerals

Fish meal: Fish meal can be produced from the fish which are not having consumers demand as fresh fish (especially marine fish) and also from the fish wastes and by-catch. Fish meal is an essential ingredient in feed formulation of fish and other domestic animals. This is mostly due to presence of growth promoting amino acids such as lysine and methionine, and other nutrients like B group vitamins, minerals like Ca, P, Na, Mg, Fe, Cu, Zn, Mn, I and Se.

Fish oil: Fish oil is obtained during wet rendering process of fish meal production from fatty fish. For human use pure quality of oil is essential, whether crude oil is having different industrial application such as in paints and varnishes, in manufacture of rubber, detergents, lubricants, printing inks, leather and cosmetics. Fish oil is also used in fish and animal feed formulation as carriers of the oil soluble vitamin A and D.

Fish silage: Fish silage is liquid fish meal produced by ensiling by-catch or fish wastes with mineral or organic acid. They are convenient ingredients in liquid feeding compounds used locally or in animal feed formulation.

Foliar spray: Foliar plant spray involves applying fertilizer directly to a plant's leaves as opposed to putting it in the soil. It is produced from fish wastes through fermentation with lactic acid bacteria.

Fish powder: Fish powder is fish protein concentrate, prepared from prime quality edible fish muscle, may be of any underutilized fish. It is a protein rich (75-85%) shelf-stable product intended for human consumption. This can be easily prepared by drying and grinding of fish mince in a controlled condition. Fish powder can also be produced through enzymatic hydrolysis of fish mince and become rich in muscle protein hydrolysates and collagen peptide. The hydrolysed fish protein powder is more suitable for infants, aged population and for the people having difficulties in protein digestion.

Fish maw and Isinglass: Fish maw is dried swim bladders that come from a large fish like a croaker or sturgeon. Usually, fish maw is used in soup and contribute as a good source of collagen, protein and other nutrients. Isinglass is also obtained from the dried swim bladder of fish, which is compressed into ribbons about 0.4mm thick. Isinglass finds applications mostly as clarifying agent for beer, cider, wine, vinegar etc.

Pearl essence: Guanine present in the scales of most pelagic fishes produces similar optical effect to that of real pearl when deposited in the surface of solid beads. Pearl essence is a suspension of crystalline guanine in water or organic solvent, and its most important use is in the preparation of artificial pearl. Also used as spray or dip for several items like shoe, pencil, fishing rod, spectacle frame, walking stick etc. to impart an iridescent sheen reminiscent of pearls and as finisher for textiles.

Fish glue: Fish glue is mainly composed of collagen and a highly viscous liquid at room temperature, produced from fish skin. It has several applications, such as in making furniture, making special cement, in photo engraving plate manufacture, book binding, etc.

Chitin and chitosan: Chitin is obtained from the wastes of shellfish through the process of demineralisation and deproteinization, and chitosan is the deacetylated form of chitin. Both chitin and chitosan have several industrial applications, which include waste water treatment, in seed coating, as growth promoting effect in broiler chickens, as gelling agent, as dietary fibre in human nutrition, in reducing cholesterol level, in would healing, in dental and neurosurgery, in reducing absorbing and binding fat, in production of contact lenses, etc. etc. Chitin is the basic materials for production of glucosamine hydrochloride, which mainly stimulates the cartilage to generate fundamental constructive material of collagen along with numerous health beneficial properties.

Collagen and gelatin: Collagen is one of the major structural components found widely in skin, bone and other connective tissues of fish and other animals. Gelatin is derived from the protein collagen, through the process of hydrolysis. Gelatin especially from warm-water fish can replace pork gelatin for use in food and drug coatings, microencapsulation of vitamins and pharmaceutical additives as well as for the production of cosmetics, photographic films and low-set glues. Several studies have shown that collagen peptides or supplements containing collagen may help slow the aging of the skin by reducing wrinkles and also give relief from osteoporosis and osteoarthritis symptoms.

Products from shark processing: Many products of commercial importance obtained during processing of shark, such as shark liver oil, shark fin rays, shark skin, shark cartilage, etc. Since these products are having export potential, therefore, may have entrepreneurial importance. Shark fin soup is a delicacy in domestic and several international markets, especially China, Taiwan and Southeast Asia. Shark skin has good export potential as this is mostly used to make leather. Shark leather is used in the production of luxury items such as handbags, shoes, cowboy boots and sandals, wallets/purses, coin/key fobs, belts, key cases, lighter cases, cigar cases, watch straps, gun holsters and knife holders. Recent interest in shark cartilage is concentrated almost entirely on its use in health supplements and as an alternative cure for certain diseases. Chondroitin is one of the active ingredients of shark cartilage which is believed to be a remedy for arthritis and ageing retardation.

Digestive enzymes: The digestive tract or the viscera, usually wasted in the markets as well as in the processing plants, are rich source of enzymes, which include pepsin, trypsin, chymotrypsin and collagenases. The fish digestive enzymes possess better catalytic properties, good efficiency at lower temperatures, lower sensitivity to substrate concentrations and greater stability in a wide range of pH. These enzymes can be commercially extracted from the fish viscera in a large scale following any low-cost method and partially purified crude enzyme may be supplied to the user for further purification. Areas of application of fish derived enzymes include pharmaceuticals, cosmeceuticals, dairy, detergent, food processing, leather etc.

As a meaningful fish waste utilization, implementation of eco-friendly recycling technologies to convert all the valuable components present in the waste into products useful for the mankind would not only reduce the amount of bio-waste going to disposal route, but would also substantially create opportunities for the entrepreneurs, resulting improvement of the economy of the fisheries sector. Products developed from fish wastes finds application in different industries such as human and animal food, pharmaceuticals, cosmetics, detergent, textile, dairy, leather, etc. Besides, many of these have immense marketable efficacy in the export market as new functional foods, nutraceuticals and dietary supplements in promoting consumer health and nutrition.

Commercial products from seaweeds

Sea weeds with their excellent nutrient value and functional compounds, are capable of meeting both domestic and international demands. There are lot of scopes for seaweed-based products for the entrepreneurs, with export potential as well as multiple benefits for coastal communities. Many useful products are made from sea weeds find application in food and pharmaceutical industries. But in India, major industrial utilization of sea weed is for production of agar and alginate, and hence, there are opportunities for the entrepreneurs.

Traditional fish products

There are number of traditionally preserved fish, which have domestic markets in many areas, especially in the northeastern states as well as export marketing potential in Southeast Asian countries. These are dried, smoked and fermented fish. An entrepreneur may go for drying fresh fish in hygienic manner followed by packaging and marketing. Alternatively, quality dry fish may be collected from the producer and marketed after attractive packaging in the retail markets. There is a good demand of smoked fish in the entire northeast region, however, smoking should be done very hygienically and the product need to be packaged with proper nutrition labelling. Most commercially important fermented fish product of northeast India is Shidal/Ngari, which is produced mainly in Tripura, Assam and Manipur. Recently, hygienic method of Shidal production has been developed. Retailer pack (pouch or jar) of 100-250g each with proper labelling will be able to capture the NEI markets. Further, due to similar food habits, Shidal is a promising product for export in the Southeast Asian markets. Production and marketing of dried, smoked and fermented fish is a good option for the entrepreneurs of northeast India.

Role of packaging in marketing of fish and products

Packaging is crucial to our modern food distribution and marketing system. The advent of modern packaging technologies and new packaging materials made possible the era of convenience products. While promoting value-added products, there is a need to develop functional, cost effective and attractive packaging distinct with various types of products. A food package usually provides a number of functions in addition to protection, which are:

- Protect the products - physical, chemical and biological agents, adulteration, tempering, contamination, damage.
- Prevent dehydration, retain flavour and colour.
- Prevent fat oxidation, rancidity.
- Help for easy distribution and handling during display.
- Serve as a communicator and provide information about product.
- Add value to the product with high quality and attractive packaging.
- Help to extend shelf-life of product.

Packaging requirements for different types of fish products

An entrepreneur should have a thorough knowledge on the packaging requirement of different fish products.

Packaging of fresh fish: A suitable package for fresh fish with ice should keep the fish moist and prevent dehydration, retard chemical and bacterial spoilage, and provide barrier against oxygen to retard oxidation of fat. Earlier and also presently in the rural areas, baskets made of split bamboo with inner lining of leaf or polythene sheet are used. But these were found ineffective in reducing bacterial spoilage and entry of external heat to melt ice. Recently, box made of high molecular weight high density polyethylene (HDPE) is found ideal even for the cyclers who sale fish door to door.

Packaging of frozen fish: Ideal characteristics for frozen fish packaging material should:

- protect the products from moisture and aroma loss
- prevent oxidation and rancidity
- not become brittle and torn
- have low water, vapour and oxygen permeability
- be able to withstand sub-zero temperatures

Most common packaging materials for the frozen fish and fish products include:

- polystyrene trays over-wrapped with polyethylene/ polypropylene film
- polyethylene bag
- plastic bag inside carton box
- waxed paper box
- interlocking, printed, polycoated and corrugated fibreboard carton

Packaging of battered and breaded products: Major issues for battered and breaded products are desiccation, discolouration and development of rancidity. Application of proper packaging prevents/retards these changes and enhances shelf life. Conventional packaging materials like flexible plastic films alone are not suitable for these products as they provide little mechanical protection to the products and as a result, the products get damaged or broken during handling and transportation. Hence, thermoformed containers are commonly used for this purpose. Trays made of materials like PVC, HIPS and HDPE are unaffected by low temperature frozen storage and provide protection to the contents during prolonged storage.

Packaging of Fish/prawn pickles: Conventionally, glass bottles are used as containers, which offer properties like inertness, non-toxicity, durability, non-permeability to gases, moisture etc. But, they are heavy, prone to break, voluminous and expensive. New flexible packaging materials developed for fish pickle is based on plain polyester laminated with LDPE-HDPE co-extruded film or Nylon/Surlyn. These are inert to the product, can be attractively fabricated as stand-up packs and printed on reverse side of the polyester film.

Extruded product: These are ready to eat snacks prepared by the extrusion process. Product is hygroscopic nature, may lead to loss of crispness of the product. Oxygen inside the package is replaced by an inert gas like nitrogen. Most important packaging requirements are low water vapour and gas permeability, physically strong enough to withstand the processes of vacuuming/gas flushing. Usually metalized polyester-polyethylene laminated pouches are used for the packaging of extruded snacks.

Packaging of dried/smoked/fermented fish products: If dried/smoked fish is not suitably packed, the fish will absorb moisture and get spoiled. So packaging is very important. The packaging material should be inert and should not allow any moisture and gas to enter the product. Dried fish are having sharp edges and spines. There is chance that they will puncture the packaging material and the moisture will enter the package. Amongst different packaging materials, high density polythene woven gusseted bags laminated with 100-gauge low density polythene are found quite suitable for bulk packaging of dried and smoked fish. These materials are cheap, readily available and have good tearing and bursting strength. Recently polyester/ polythene laminated pouches and PET jars are in use for packaging of fermented fish products.

Chitin and chitosan: Bulk packaging is done in HDPE woven gusseted bag laminated with 100-gauge LDPE liner.

Fish meal: Packaging should be impermeable to moisture, oxygen and other insets and pests. Generally packed in HDPE sacks for bulk transportation. Fish meal is also packed in jute bags, multiwall paper bag which are lined with polythene and in HDPE woven bags with liner.

Fish oil: Easily susceptible to oxidation when exposed to air. Packaging material needs high barrier properties, moisture proof, oil resistant and impermeable to oxygen. Larger quantities of fish oil are mainly packed in Linear low-density polyethylene (LLDPE)/Nylon films or in glass bottles.

The retail pack must be clear, attractive and cheap. The contents of the pack should be clearly mentioned on the pack in an attractive way. When a new product is introduced, the package should carry information like instructions to prepare product, caloric value, nutritional information, dietary plans and health benefit to give confidence to the consumer.

Labelling requirement as per Food Safety and Standards (Packaging and Labelling) Regulations, 2011

Labelling should provide the information of:

- Name of food
- List of ingredients
- Nutritional information

- Declaration regarding food additives
- Name and complete address of the manufacturer
- Net quantity
- Lot/Code/Batch identification
- Best Before and Use by Date
- Country of origin for imported food
- Instructions for use
- Mandatory use of logo of 'Non-veg' food

SSOP (Sanitation Standard Operating Procedures) in production centre and quality control

There are many challenges specifically in issues concerning food safety and quality which need to know by the entrepreneurs and must be addressed. Delivery of quality and safe fish and fish products will attract the confidence of the consumers to a great extent. Flesh of live fish is free from any bacteria although they remain present in the skin, gills and in the intestine. After death, all the organisms present in the skin, gills and in the intestine compete to consume fish and fish products. If handled carelessly fish and fish products can become a source of different kinds of public health organisms, i.e. pathogenic bacteria. Elevated temperature, poor sanitary conditions of the factory premises and inadequate hygiene of the workers will speed up the fish spoilage and contamination. The investment in the fish processing unit will result only in diminishing returns unless a well-organized programme of sanitation is planned and effectively implemented. Some guidance which are important for fish processing unit to maintain SSOP and quality of the fish products are discussed here.

Care on handling of fish

- Only fresh raw materials, i.e., fin fish and shell fish has to be procured.
- Immediately after harvesting or procurement of fish they should be washed and shall be mixed with ice in the ratio 1:1 and stored in clean containers till its use.
- Fish should be protected from flies, rodents, insects, birds and animals.
- All fish handling surfaces. shall be smooth, non-corrosive and free from cracks and crevices.

- Bamboo baskets, wooden boxes and similar containers should be avoided as those are difficult to be cleaned.
- Fish handling area and containers should be properly disinfected.
- The boxes used for keeping fish shall be of plastic/LDPE or any other suitable synthetic material that can be easily cleaned.
- Ice should be used always during transportation of fish.
- Care must be taken so that the fish is not damaged physically while handling as this will speed up the rate of spoilage.
- Shrimps should be graded, beheaded, peeled and de-veined as soon as possible.
- The bivalves as far as possible should be depurated before shucking.
- Sun drying of fish in sandy beach should be strictly stopped.
- While salting, only good quality fish, salt and clean containers should be used.

Facilities in the processing plant

- An adequate supply of potable water (drinking water quality) with appropriate facilities for its storage, distribution should be available whenever necessary to ensure the safety of fish.
- Adequate facilities should be there for chlorination of water intended for different uses where disinfection is required.
- Adequate drainage and waste disposal systems and facilities should be provided. They should be designed and constructed so that the risk of contaminating fish or the potable water supply is prevented.
- Adequate cleaning facilities should be suitably designated and provided for cleaning fish, utensils and equipment. Such facilities should have an adequate supply of hot and cold potable water where appropriate.
- Depending on the nature of the fish operations undertaken, adequate facilities should be available for heating, cooling, cooking, refrigeration and freezing fish, for storing refrigerated or frozen fish, monitoring fish temperatures to ensure the safety and quality of fish.
- Adequate means of natural and mechanical ventilation should be provided, in particular to:

- minimize air born contamination of fish from aerosols and condensation droplets
- control ambient temperatures
- control odours which might affect the suitability of fish
- control humidity, where necessary to ensure the safety and suitability of fish.

- Ventilation systems should be designed and constructed so that air does not flow from contaminated area to clean areas and where necessary, they can be adequately maintained and cleaned.
- Adequate natural and artificial lighting should be provided to enable processing operations in a hygienic manner.
- Lighting fixtures should, where appropriate, be protected to ensure that fish is not contaminated by breakages.
- Where necessary, adequate facilities for the storage of fish products, ingredients and non-food chemicals (e.g., cleaning materials, lubricants, fuels) should be provided.
- Storage facilities of fish products should be designed and constructed to:
    - permit adequate maintenance and cleaning
    - avoid pest access and harbourage
    - enable fish products to be effectively protected from contamination during storage
    - when necessary, provide an environment which minimize the deterioration of fish products (e.g., by temperature and humidity control).

Cleaning procedure will involve, where appropriate

- to remove debris from surfaces
- to apply a detergent solution to loosen soil and bacterial film and hold them in solution or suspension
- to rinse with water to remove loose soil and residues of detergent
- for dry cleaning or other appropriate methods for removing and collecting residue and debris

Personal hygiene

- People known or suspected to be suffering from or to be a carrier of a disease or illness should not be allowed to enter any fish handling areas if there is likelihood of their contaminating fish and fish products.
- Suitable waterproof dressing should cover cuts and wounds, when personnel are permitted to continue working.
- Food handlers should maintain a high degree of personal cleanliness and where appropriate, wear suitable protective clothing, head covering, face mask and footwear.
- Personnel should always wash their hands at the start of food handling activities, immediate after using the toilet; and after handling raw food or any contaminated material.
- People engaged in food handling activities should refrain from behaviour, which could result in contamination of food, for example: smoking, spitting, chewing or eating, sneezing or coughing over unprotected food.
- Personal effects such as jewellery, watches, or other items should not be worn or brought into food handling areas if they pose a threat to the safety and suitability of food product.
- All personnel should be aware of their role and responsibility in protecting food from contamination or deterioration.
- Managers and supervisors of food processes should have the necessary knowledge of food hygiene principles and practices to be able to judge potential risk and take a necessary action to remedy deficiencies.
- Visitors to food manufacturing, processing or handling areas should wear appropriate protective clothing and adhere to the other personal hygiene provision given in the above.
- Personal hygiene facilities should include:
    - adequate means of hygienically washing and drying hands including wash basins and supply of hot and cold water
    - lavatories of appropriate hygienic design
    - adequate changing facilities for personnel.

Role of Government

Government should play a catalyst role to support the entrepreneurs to sustain and grow in the market with their innovative ideas and products. The focusing points of the Governmant should be as follows:

- Modernization of the fish landing centres and fishing harbours with all necessary accessories such as ice plant, potable water, refrigerated vehicles, auctioning facilities, storage, etc.
- Creation of hygienic fish markets and fish processing parks
- Creation of cold storage ware houses
- Creating consumer awareness on nutritional benefits of consuming better-quality fish
- Creation of an atmosphere conducive for providing access and / or control over land and basic infrastructure
- Provision of affordable finance, support systems, specific credit and market support services
- Creation of facilities for skill up-gradation including training and education
- Promotion of linkages for adequate finance, linking fish and fishery products to domestic and export market etc.
- Creation of 'Incubation Centre' at zonal level for skill development and hand-holding of the potential entrepreneurs. A successful business incubator can create a good entrepreneurial environment and greatly improve a new enterprise"s chance for success.
- Creation of a cordial atmosphere between government and potential entrepreneurs to avail the benefits of different govt. financial schemes

Role of a business incubator

Business incubators play a key role in translating an innovative idea of an entrepreneur to reality. A successful business incubator can create a good entrepreneurial environment and greatly improve a new entrepreneur's chance for success. A business incubator plays a significant role for imparting pro-active, value-added support in terms of scientific and technical consultancy, access to critical tools such as entrepreneur-ready technologies, vast infrastructure and other resources to create a sustainable market for well-tested technologies. An ideal business incubation centre creates a friendly eco-system to start the business with the following facilities:

- Constant mentoring
- Upgradation of skills and techniques
- Technology optimization and upscaling
- Preparation of business plan and marketing consultancy
- Access to funding agencies, angel investors, venture capitalists, crowd-sourcing, etc.
- Branding and labelling
- IP protection
- Promotional support through exhibitions, conventions, workshops, etc.
- Facilitate certifications/ approvals/ registrations
- Linkages to industry and other business mentors
- Facilitate e-marketing
- Assisting in getting clearances with government regulatory bodies and Licensing bodies such as FSSAI, Pollution Control Board, etc.
- Establishing a network, etc.

Following business opportunities need to be available with the business incubators:

- Fish pre-processing line
- Retort pouch processing unit
- Fish canning line
- Fish sausage production line
- IQF product line
- Fish extruded product line
- Fish curing and drying line
- Fish battering and breading line
- Fish product packing system line
- Chitin and chitosan production line
- Fish meal/fish oil production line, etc.
- Valued products from fish wastes

Conclusion

Technology developments in fish processing sector offer scope for innovation, increase in productivity, increase in shelf life, value addition, product diversification, improve food safety and reduce waste during processing operations. However, the commercialisation of fish products still poses lot of challenges to the entrepreneur and researcher in terms of optimization of technologies and ultimately developing the technologies into a commercially viable business plan. Functional fish products will be in much demand in future; the challenge will be to retain the functional benefits of fish and shellfish meat by way of adopting product specific processing protocols or alternate delivery systems for sensitive components. These issues offer ample scope for innovation coupled with entrepreneurial skills for the creation of wealth and employment in fisheries sector. Presently, entrepreneurship is set to play a vital role in leading the Indian economy in the future. With its ability to create jobs, spur innovation, and promote economic growth, entrepreneurship will continue to be a key driver of the Indian economy for years to come. The government's support for entrepreneurship is a positive step in the right direction, and it will be interesting to see how entrepreneurship continues to shape the Indian economy in the years to come.

References

Bindu J, Sreejith S. 2018. Packaging of Seafood products, pp 53-60. Training Manual on Seafood Value Addition, ICAR-Central Institute of Fisheries Technology.

Cold Chain and Post Harvest Infrastructure Development, 2022. Department of Fisheries, Ministry of Fisheries, Animal Husbandry & Dairying, Govt. of India, New Delhi.

Draft National Fisheries Policy. 2020. Department of Fisheries, Ministry of Fisheries, Animal Husbandry & Dairying, Govt. of India, New Delhi.

Majumder RK. 2019. Advances in Fish Processing Technology, pp 163-190. Technologies for Value Addition in Food Products and Processes, (Eds) Deka SC, Seth D and Hulle NRS. Apple Academic Press Inc., Taylor & Francis Group, USA.

Majumder RK. 2019. Traditional Technology of Fish Preservation in Northeast India, pp 127-162. Technologies for Value Addition in Food Products and Processes, (Eds) Deka SC, Seth D and Hulle NRS. Apple Academic Press Inc., Taylor & Francis Group, USA.

Majumder RK. 2023. Fish Wastes - Present Status of Utilization and Scope of Valorization, pp 289-310. Advances in Fish Processing Technologies- Preservation, Waste Utilization, and Safety Assurance, (Eds) Majumder RK, Balange AK. Apple Academic Press, USA, CRC Press, Taylor & Francis Group.

Martin X, Chattopadhyay K. 2023. Bioactive and functional nutraceuticals from shrimp processing wastes, pp 321-336. Advances in Fish Processing Technologies- Preservation, Waste Utilization, and Safety Assurance, (Eds) Majumder RK, Balange AK. Apple Academic Press, USA, CRC Press, Taylor & Francis Group.

Murthy LN, Phadke GG, Jeyakumari A and Vijayakumar S. 2023. Secondary Raw Material and Low-Cost Fish for Fish Meal Production and its Implications in Animal Feed Industry, pp 183-200. Advances in Fish Processing Technologies- Preservation, Waste Utilization, and Safety Assurance, (Eds) Majumder RK, Balange AK. Apple Academic Press, USA, CRC Press, Taylor & Francis Group.

Ninan G. 2023. Entrepreneurship Initiatives in Fisheries Post Harvest Sector through the Agri Business Incubation Programme, pp 453-492. Advances in Fish Processing Technologies- Preservation, Waste Utilization, and Safety Assurance, (Eds) Majumder RK, Balange AK. Apple Academic Press, USA, CRC Press, Taylor & Francis Group.

7

Entrepreneurship Opportunities in Ornamental Fish and Aquarium Industry of India

Atul Kumar Jain

*Ornamental Fisheries Training and Research Institute, Udaipur-313001 Rajasthan (*www.oftri.org*)*

Abstract

The hobby of ornamental fish keeping is gradually becoming very popular in India. About 1.25% of urban house-holds in country were estimated keeping an aquarium either at home or work place during 2015-16. It is expected to grow to 5.0% in the present decade. It will be possible as both annual average per capita income (AAPCI) and number of Middle-Income Group House-holds (MIGHh); the two important economy indicators of a developing nation are continuously increasing. The AAPCI which was INR 53331 only in year 2011 increased to INR 77826 in year 2015-16 and estimated to increase to INR 290000 in year 2026. The number of MIGHh has increase from 31.4 million in year 2011 to 50.0 million in year 2015-16 and estimated to increase to 113.80 million in year 2025-26. The domestic aquarium trade increased from Rs. 55 crores at the end of year 2000 to Rs. 555 crores by 2019-20. The Government of India had now recognized the potential of ornamental fisheries for generating entrepreneurship and providing livelihood to youths of the country. Therefore, it is one major focus area under Pradhan Mantri Matsya Sampada Yojna (PMMSY) launched in 2019-20 with proposed investment of Rs. 576 crores for ornamental fisheries development, targeting domestic aquarium trade value of Rs. 3000 crores by the year 2025. The domestic aquarium trade is shifting from a low-profile aquarium shop to a very highly modernized stores that are being managed by experienced and qualified fisheries professionals. Several public aquarium galleries have also come in existence during the last 8-10 years in several cities of the country, latest and largest being Aquatic Gallery at Science City Ahmedabad. The ornamental fishery is presently not limited to the traditional practices of ornamental fish production in a backyard facility and setting up of an aquarium shop. The scope of entrepreneurship in ornamental fisheries has become very vast presently which is discussed at length in the present chapter.

Keywords: Ornamental fish, Domestic aquarium trade, PMMSY, Fish pedicure spa, Public aquarium gallery, Aquarium plant, Live feed culture

Introduction

The development of ornamental fish and aquarium industry in India could be divided into three phases i.e., pre-juvenile, juvenile and post juvenile phase. The period, since independence till the beginning of the 21st century was the pre-juvenile phase of ornamental fish and aquarium industry. It was characterised by few numbers of aquarium shops mainly in metro and major cities of the country, availability of limited varieties of fish and aquarium products, high cost of products, very few number of hobbyists and lack of any institutional support. The ornamental fish and aquarium industry during this period had developed because of the individual efforts of ornamental fish producers and aquarium traders. The total value of domestic aquarium trade was Rs. 55 crores, the export of ornamental fish from country was Rs. 1.74 crores (US$ 0.40 million) during 1999-2000 and aquarium keeping was known to be the hobby of rich only.

The span of juvenile phase was from the beginning of the 21st century to the year 2019-20. The hobby of aquarium keeping grew significantly during this period as about 1.25% of the urban house-hold were estimated to keep an aquarium at home or workplace. It was because availability of aquarium and aquarium accessories increased in the domestic market due to entry of Chinese products, the cost of fish and aquarium products reduced, the number of aquarium shops increased with existence even up to small cities and most importantly the availability of institutional support. The Marine Products Export Development Authority (MPEDA), Ministry of Commerce and Industry, Government of India, started providing financial assistance from 2006-07 to beneficiaries to set up ornamental fish production units. The schemes of MPEDA initially covered the coastal states of Karnataka, Kerala, Maharashtra, Tamil Nadu, and West Bengal and later inland states of Himachal Pradesh, Madhya Pradesh, and Rajasthan. A total of 579 ornamental fish production units of different grades were established in all these states but the scheme was closed in the year 2014-15. It was followed by another pilot project implemented by National Fisheries Development Board (NFDB), Ministry of Fisheries, Animal husbandry and Dairying, Government of India (GOI) with a proposed investment of Rs. 61.98 crores. It covered six coastal states namely Gujarat, Karnataka, Kerala, Maharashtra, Tamil Nadu, and West Bengal, an inland state Bihar and one hilly state of Assam from north-east. The total value of domestic aquarium trade was about Rs. 555 crores during 2019-20 and the export of ornamental fish from country was Rs. 10.84 crores. The aquarium keeping was known to be the hobby of common man now.

The post-juvenile phase started after the implementation of Pradhan Mantri Matsya Sampda Yojna (PMMSY) by GOI in year 2019-20. The ornamental fisheries was identified as one of the thrust area for generating employment and livelihood opportunities for the youth of the country. An investment of Rs, 576 crores was proposed for development of the sector with an additional financial support of Rs. 500 crores from the world bank. The domestic aquarium trade was targeted to achieve a value of Rs. 3000 crores and export to Rs. 100 crores by the end of year 2025. The PMMSY is implemented across the country and more than 1800 ornamental fish production are already established in different states of the country since its implementation.

All the above stated growth trends, economic growth indices viz; increasing per capita annual income, increasing number of middle-income group families, increased human development index in addition to increasing urbanisation, increasing number of nuclear families etc. indicates the potential of developing ornamental fisheries and aquarium industry in India (Table-1).

Table 7.1: Development of Indian aquarium industry during different phases

S.No.	Growth Indicator	Development Phases of Aquarium Industry		
		Pre-Juvenile	**Juvenile**	**Post-Juvenile**
1.	Period	Independence to the beginning of 21st century	Beginning of 21st century to 2019-20	2019-20 onwards
2.	Value of Trade	Rs. 55 crores	Rs. 555 crores	Rs. 3000 crores*
3.	Major production Centres	West Bengal and Maharashtra	West Bengal, Kerala, Tamil Nadu, Karnataka, and Maharashtra	Throughout the country
4.	Distribution of shops	Mostly metro cities and Mega cities	Metro cities, Mega cities, almost all district HQs	Increased numbers in metro and mega cities, almost all district HQs and even in large towns.
5.	Number of aquarium shops	-	5000	Above 8000
6.	Cost of products	++++	+++	+++
7.	Cost of fish	+++++	++++	+++
8.	Varieties of fishes	++	++++	++++++
9.	Hobbyists' choice of fish	Value based (Low to medium value)	Value based (Low to high value	Variety based irrespective of value

10.	Number of hobbyists	-	1.25% of urban house-holds	5% of urban house-holds
11.	Export (In crores)	Rs. 1.74 (1999-2000)	Rs. 10.84 (2019-20)	Rs. 100.00 End of 2025
12.	Institutional support as proposed investment In crores)	NIL	**2006-07 to 2014-15** MPEDA: Rs. 19.15# NFDB: Rs. 31.15 # **2014-15 to 2019-20** NFDB: Rs. 61.89	PMMSY: Rs. 576 World Bank: Rs 500
* Projected at the end of PMMSY by 2025; # Actual Expenditure				

Now, the ornamental fisheries do not remain limited to culture of ornamental fishes and setting up of an aquarium but it includes number of activities for entrepreneurship development (Fig.1). The existing growth rate of Indian aquarium industry coupled with continuing economic development indicates that the number of aquarium hobbyists in India will be much more than the several developed countries by the end of present decade. The Indian aquarium industry offers vast opportunities for entrepreneurship development to the farmers, educated & uneducated youth, women, weaker sections of the society and many others in both rural and urban regions.

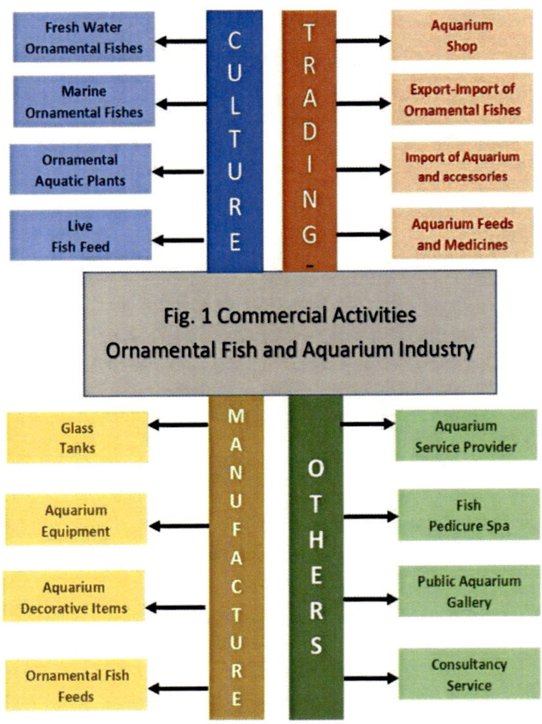

Fig. 1 Commercial Activities Ornamental Fish and Aquarium Industry

1. Culture: It includes farming of freshwater ornamental fishes, marine ornamental fishes, ornamental aquatic plants, live fish feed culture etc.

1.1 Freshwater ornamental fishes

There are more than 500 varieties of freshwater ornamental fishes that are found in domestic aquarium trade. The most popular varieties are Gold fish, Barbs, Tetra, Gourami, Angels, and Live bearers. These are mainly the low and medium value fishes which are in high demand in the domestic aquarium trade but gradually the demand of high and premium value fishes is also growing (Table-2). The coastal state of West Bengal, and Tamil Nadu are the major centres of freshwater ornamental fish production in the country followed by Kerala, Karnataka, and Maharashtra.

Strengths & Opportunities

- A backyard unit can be established in an area of 1000 sq. ft.
- The activity can be started with an initial investment of Rs. 1.00 lakh or less.
- Easy to learn breeding and culture technology.
- Ready market for healthy and high-quality fishes.

Table 7.2: Commonly traded freshwater ornamental fishes in domestic market

Category*	Name of Varieties (Common Names)
Low value (Rs.15-50 each) #	Varieties of Gold fish, Koi carps (Golden, Milky), Sharks (Tiger, Albino), Gourami (Blue, Kissing, Golden), Angels (Black, Tiger, Golden), Barbs (Rosy, Tiger, Tin foil),Tetra (B.A.,Widow, Serpae), Live Bearers (Mollies, Sword tails, Guppy, Platies), Fighters etc.
Medium value (Rs.50-200 each) #	Varieties of Gold fish, Koi carps (Golden, Milky), Sharks (Silver, Redtail, Rainbow), Gouramy (Pearl, Giant), Angels (Platinum, Diamon, Koi), Barbs (Filamentosa, Denisonii, Ticto), Tetra (Neon, Cardinal, Rummynose), Live Bearers (Baloon Mollies, Sword tails, Guppy, Fighters, Oscars, Silver Dollars, Fire mouth, Shevrum, Feather fish, shrimps
Hing value (Rs. 200-2000 each) #	Varieties of Gold fish, Koi carps (Golden, Milky), Oscars, Silver Dollars, Parrot, Texas, Flower Horns, Varieties of Discus, Red devil, Green terror, Lobsters, Several varieties of cichlids, Marine ornamental fishes
Premium value (> Rs. 2000 each) #	Flower Horn, Discus, Arowana, Marine ornamental fishes
** Few fishes are grouped in more than one category based on size, variety, colour etc.* # Maximum retail price	

The farming of ornamental fishes is commonly practiced in coastal states as environment temperature remains constant throughout the year due to absence of winter months. But the agro-climatic conditions of inland states are also suitable for ornamental fish breeding and culture except during the winter months which can be overcome by setting up of a polyhouse or in a controlled environment. A few numbers of farms have been established in the inland states of Rajasthan, Madhya Pradesh, and Uttar Pradesh but it is mainly West Bengal and Tamil Nadu supplying fish to all the other states of the country. The fishes are mainly transported through train as air freight charges are very high while the travel time in train is very long which puts the fishes in stress and hobbyists do not get healthy fishes. Therefore, aquarium fish retailers are always looking for ornamental fishes produced in the vicinity and there is good demand of fishes every time. An entrepreneur can set up a backyard unit (Picture-1), middle scale unit or a high-tech unit (Picture-2) depending on the availability of resources, and technical expertise available.

Fig 7.1: A back yard ornamental fish production unit at Ernakulum

Fig. 7.2: A hi-tech ornamental fish production unit at Vijayawada

Threats & weaknesses: A beginner is advised to start with the breeding and culture of low or medium value fishes in a backyard unit and gradually upgrade to high value or premium value fishes and setting up of high-tech system. There may be difficulties with existing old producers in term of price.

1.2 Marine ornamental fishes

The keeping of marine ornamental fishes is also gaining popularity specifically in metro and coastal cities (Picture-3). The installation and maintenance of marine aquarium is more technical and 8-10 times costlier than a freshwater aquarium. The breeding technology of many varieties of ornamental fishes viz; clown fishes (Picture-4) and Damsels has been developed at ICAR-Central Marine Fisheries Research Institute, Kochi, and Annamalai University. But commercial scale breeding of marine ornamental fishes has yet not started in the country. The marine fishes sold in the domestic trade are either of wild collection or imported from other countries. The wild collection of marine ornamental fishes specifically those inhabiting the coral reefs are not permissible. Therefore, the setting up of a large-scale marine ornamental fish production unit could be another highly profitable venture.

Fig 7.3: A marine aquarium set up

Fig. 7.4: A popular marine aquarium fish Amphiprion percula

Strengths and Opportunities

- The hobby of keeping marine ornamental fishes and invertebrates is growing.
- High market demand as there are not many commercial production units.
- Breeding technology of commercially important species is available.
- Funding support available under PMMSY

Threats and weaknesses: A highly technical activity with very high initial investment requirement. It may be difficult to compete with illegally collected wild stock as well as imported fish.

1.3 Aquatic ornamental plants

Many hobbyists prefer to keep a planted aquarium either for their liking for ornamental aquatic plants or to keep some specific varieties of fish that look beautiful in a planted tank (Picture-5). Some hobbyists are interested to keep only planted aquarium and not fish due to religious or some other reasons. As a result, the trend of keeping a planted aquarium is becoming very popular specifically in the cities and the demand of aquatic ornamental plants as well as other components required to set up a planted tank is constantly increasing. The setting up and maintenance of planted aquarium is quite different than a fish tank. There are not many ornamental aquatics plant cultivation units in India like other south-east Asian countries. The plants can be cultured in earthen ponds or cement tanks (Picture-6). Now days, tissue cultured ornamental aquatic plants are also available in the market. Many of the common varieties viz *Vallisneria, Bacopa, Amazon, Hydrilla, Cabomba* etc. are also collected from ponds, tanks, and other water bodies while exotic varieties are imported. The demand of aquatic ornamental plants is regularly increasing but there is scarcity of good plant producers and suppliers. Financial assistance is also available under PMMSY to set up an aquatic plant culture unit.

Fig. 7.5: An ornamental aquatic plants aquarium setup

Fig. 7.6: An ornamental aquatic plant culture farm

Strengths and opportunities

- A fast-growing hobby of keeping planted aquarium.
- Outdoor or Indoor units can be established in type of climatic condition.
- Only few aquatic plant farming units exists in the country presently.
- Funding support available under PMMSY

Threats and weaknesses: The cost of wild collected plants is very less compared to cultured plants so preferred by aquarium traders and presently the demand is mainly restricted to metro and large cities.

1.4 Breeding of native fish species

There are many varieties of colourful fishes found in the rivers, reservoirs, lakes, and stream of the country. The Wester Ghats and north-east states of the country are known as two important hot spots and rich resources of native freshwater ornamental fishes. There is not much demand of these fishes in the domestic aquarium trade in the global aquarium trade. About 85-90% of the fish exported from India are from the Western Ghats and north-east states. The stocks of many of these fish species are declining in the nature for some reason or other and environment conservationist blame aquarium trade for their depletion. However, the demand of these fishes is continuously growing even in the domestic market with the popularisation of planted aquarium. The successful captive breeding technology of several native species viz; *Ctenops nobilis* (Picture-7), *Garra annandalei*, *Mirophis deocata*, *Danio dangila*,

Esomus danricus, Sahayadria denisonii and many more has been successfully developed in the country. Therefore, the setting up of commercial facilities of producing native ornamental fishes could be a very flourishing business activity.

Fig. 7.7: Ctenops nobilis sold for 8-10 US$ each in international aquarium trade

Strengths and opportunities

- High demand of native freshwater ornamental fishes of India in global aquarium trade.
- Demand is gradually growing in domestic trade with increasing demand of planted aquarium.
- Breeding technology is developed for many of the species.

Threats and weaknesses: The native fish species collected from natural water bodies are more vibrant and cheaper hence preferred by exporters. The demand in domestic aquarium trade is not much at present.

1.5 Live fish feed culture

Many types of small organisms found in nature are eaten by fishes as food. These are collectively called as live fish feed because eaten by fish in living form (Picture-8). It includes Daphnia (Zooplankton), Moina (Zooplankton),

Blood worms (Larvae of *Chironomus* fly), Tubifex worms (Annelids), Artemia (Crustacean) etc. Live fish feeds are highly nutritive and balanced so considered excellent for brood stock development, rearing of young ones and colour enhancement of ornamental fishes. Many aquarium hobbyists also prefer to feed their fish on these live feeds to maintain health and colour of fishes. All these live feeds are mainly collected from natural sources and supplied to users. It is a very flourishing business activity around all the major ornamental fish production centres in the country. However, the use of such live feed is not advisable as it is mainly collected form sewage channels. It is full of external and internal pathogens that lead to disease outbreak at production centres, aquarium fish dealers shop and in the aquarium of hobbyists. The commercial culture of all these live feed organisms could be a highly profitable business activity. The live feeds can be cultured either in earthen ponds, FRP tanks or cemented tanks (Picture-9) depending on the availability of space. It can be sold both fresh or in freeze-dried form.

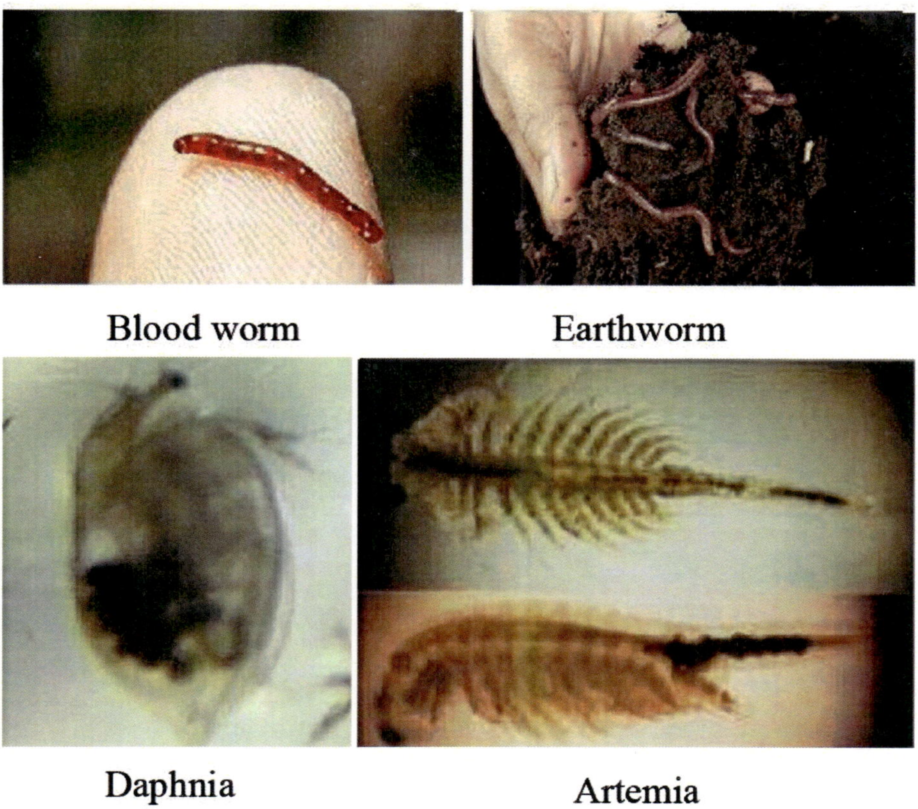

Fig. 7.8: Different types of live fish feeds

Fig. 7.9: A small Tubifex culture unit

Strengths and opportunities

- Ever increasing demand among ornamental fish producer and aquarium hobbyists.
- Commercial scale production technologies are available.
- Can be cultured indoor or outdoor throughout the year

Threats and weakness: Only large-scale operations may be economically viable. A regular and planned harvest is essential to obtain high yields else the culture may crash.

Strengths and opportunities

- Ever increasing demand among ornamental fish producer and aquarium hobbyists.
- Commercial scale production technologies are available.
- Can be cultured indoor or outdoor throughout the year.

2. Trading: The trading activities related to ornamental fish and aquarium trade includes; setting up of a wholesale or retail aquarium shop, export-import of ornamental fishes, import of aquarium and aquarium accessories etc.

2.1 Setting up of an aquarium shop

The aquarium shops were mainly limited to metro and major cities of the country prior to the beginning of 21st century. As the availability of aquarium and aquarium accessories increased with introduction of Chinese products which were of superior quality and cheaper than domestic products, the number of hobbyists started increasing. It resulted in increased demand and a greater number of aquarium shops. There were about 5000 aquarium shops in the country during 2015-16 with only 1.25% urban house-holds keeping aquarium. The number of aquarium shops increased to 8000 by the end of year 2021 and the demand is increasing continuously. It was estimated that there is demand of about 28000 aquarium shops in the country if 5% urban house-hold will be keeping an aquarium. The setting up and managing of a retail aquarium shop does not remain a low-profile business activity as it was two decades back. It has now become a very lucrative business activity being established by qualified fisheries professional and highly educated youths from engineering, science, and MBA qualification. A basic set up of an aquarium shop can be established with an investment of Rs 1.0-2.0 lakhs only and the venture becomes economically viable with a customer base of about 200 aquarium hobbyists.

Fig. 7.10: A new generation aquarium shop at Udaipur

Strengths and opportunities

- Demand of aquarium shops is growing with ever increasing number of hobbyists.
- Can be started as a home-based activity initially and gradually upgraded to a proper shop.
- Specialised aquarium setup may result in high profitability.

Threats and weakness: The profit margins are declining with ever increasing numbers of aquarium shops. A low-profile set up without adequate technical skills may not be accepted by hobbyists.

2.2 Export of ornamental fishes

The global export of ornamental fishes was valued US$ 322 million during 2019-20. However, the contribution of India was only US$ 1.55 million (Rs. 10.84 crores i.e., 0.48% only. The export of ornamental fishes from India started since the year 1969 but the situation is not changed. The major player in global ornamental fish export are mainly Asian countries only viz; Japan, Singapore, Indonesia, Sri Lanka and Malaysia which are much smaller than India in terms of geographical area and fisheries resources. India is known in international trade only for wild collected species while most of the other countries for captive bred species including native species of India. The country could not make a dominant presence in international ornamental fish trade due to low acceptance of our fishes by major importing countries. A one of the important reasons for the same is lack of large-scale ornamental fish production units with adequate fish holding and quarantine facilities so to produce high quality fishes to cater to international trade (Picture-11). Therefore, setting up of an "Export oriented unit" of freshwater/marine ornamental fishes/ornamental aquatic plants could be a highly profitable but challenging entrepreneurship development activity.

Fig. 7.11: An export-import ornamental fish unit at Singapore

Strengths and opportunities

- Demand of many native species of India in global aquarium trade.
- The export from India have increased significantly post Covid-19.

- Breeding technology of many native species is developed.
- Financial assistance available under PMMSY to set up export-oriented unit.

Threats and weaknesses: The contribution of India could not increase above 0.5% in global ornamental fish export since the beginning in 1969. The small units are not viable and the cost of large units is very high. It may be difficult to compete with established exporters.

3. Manufacturing: There are many manufacturing-based activities related to ornamental fish and aquarium trade viz; manufacturing of aquarium, aquarium equipment, decorative items, fish feed, medicines etc.

3.1 Aquarium and aquarium accessories

The glass tank which is used as aquarium, essential aquarium equipment which are used for maintaining the water quality and the artifacts which are used for aquarium decoration contribute about 63% to the total value of domestic aquarium trade (Picture-12). Before the year 1990 most of these products were being manufactured in India but after economic liberalisation policy of GOI post 1990, Chinese aquarium products entered in domestic aquarium trade. All these products were of very good quality, available with varieties and less priced than the Indian products hence, easily replaced the local products in a short span of time. It will not be untrue to say that it was because of these Chinese aquarium products, that the hobby of aquarium keeping started growing and achieved to this level. In the present scenario, 95% of the aquarium and aquarium accessories sold in the domestic aquarium trade are of Chinese origin. However, the situation is changed post COVID-19. The cost of Chinese aquarium products has gone up, the GOI has increased import duty and the entire world countries are looking for an alternative of Chinese products and eyeing at India.

Now, there is immense scope to set up manufacturing units of all these items (Table-3) indigenously owing to present technological advancements in the country. The financial support to set up these units could be availed from the Ministry of micro, small & medium enterprises GOI as well as under Fisheries Infrastructure Development Fund of Ministry of Fisheries, Animal Husbandry and Dairying.

Moulded Glass Tank

Air pump

Filter pump

FRP toys

Plastic plants

Fig. 7.12: Aquarium tank, accessories, and decorative items

Strength and opportunities

- Indigenously manufactured superior quality products are not available in the market.
- India has technically advanced in manufacturing technology.
- Demand in domestic and global aquarium trade.

Precautions: The initial investment of a manufacturing unit will be very high and there will be a tough competition with Chinese products that are already established in the market.

Table 7.3: Types of industries for manufacturing aquarium products

S.No.	Type of Industry	ITEMS	Remarks
1.	Glass	Glass tanks of various sizes and shapes	Fabrication of glass tanks with molded front in different shapes (C, U) and dimensions.
2.	Acrylics	Acrylic panels of large sizes and thickness for large aquarium tanks and public aquarium galleries.	The high import cost of large size acrylic panels is a major limitation to take up mega public aquarium gallery projects in country.
3.	Electricals	Air pumps, Compressors, blowers, Thermostatic heaters, Filter pumps, Lights etc.	Very high demand on day-to-day basis. A few of these products were being manufactured indigenously but could not compete with Chinese products in terms of price, quality, and variants hence all the units were closed.
4.	Plastic	Parts for air pumps and filters, Aquarium covers and base, Artificial aquatic plants, Aquarium toys, Aeration tubes, Joints, Regulators, Siphoning pipes, Glass wipers, Magnetic cleaners etc.	Very high demand on day-to-day basis. Manufacturing of good quality products may generate very high demand.
5.	Printing	Water proof background posters, Packaging for aquarium and aquarium accessories.	Background posters are laminated sheets which are in high demand but not manufactured indigenously.
6.	FRP	Aquarium decorative items.	Only few local manufacturing units are functional but mainly imported due to high cost and a smaller number of new designs.
7.	Metal	Aquarium cover & frames.	It will be an integral part of glass tanks.
8.	Wooden	Aquarium stands	The availability of folding structure and easy assemblage of the units has created very high demand while the quality of imported product is inferior.

3.2 Manufacturing of feeds and medicines

The feed and medicines contribute about 7% to total value of domestic trade. The ornamental fish feed industry has grown very well overseas with several internationally established brands viz; Serra, Taiyo, Hikari, Tetrabits, Optimum, Boltz etc. These companies are manufacturing species specific fish feed which are in high demand in the domestic market (Picture-13). A few of the entrepreneurs in India are importing feed of these brands in bulk and marketing locally in small packings. A few of the edible fish feed manufactures in India have introduced ornamental fish feeds also during last 1-2 years with good success. But the Indian domestic market is very large and still there is scope for high quality ornamental fish feeds. The similar is the situation with medicines required for ornamental fish disease treatment and water quality treatment chemicals. A few of the basic products are manufactured indigenously but high value products are of exotic origin only.

Fig. 7.13: Commercial brands of ornamental fish feeds

Strengths and opportunities

- Most of the ornamental fish feeds and medicines sold in domestic trade are of exotic origin.
- Hobbyists prefer to pay any amount for good quality feeds.
- Funding support available under PMMSY

Threats and weaknesses: Acceptance of any new brands of feed by hobbyists is very difficult. Small scale production may not be economically viable. A few of the large manufacturers could not be successful with new brands of ornamental fish feeds.

4. Others

4.1 Aquarium servicing and maintenance

The keeping of aquarium both at home and commercial centres have increased several folds during last two decades. A hard-core hobbyist prefers to service the aquarium at his own but many of the hobbyists are not able to do routine cleaning or water exchange of the aquarium. It is more difficult at commercial establishment due to common ownership, large size of the aquarium and lack of technical knowhow. A few of the working house-holds avoid to keep aquarium for the problem of routine cleaning. Hence, there is good demand of aquarium service providers in the cities. The job of aquarium servicing is not highly technical and can be learned easily with few months of job experience at an aquarium shop and small investment of Rs. 5000/- on a service kit (Table-4). An aquarium service providers could become a certified "Aquarium Technician" after attending an off-line or on-line training program conducted by an institution recognised by National Skill Development Corporation, Ministry of Skill development and entrepreneurship, GOI.

Table 4: Items in an aquarium servicing kit

S.No.	Requirement	Name of Item
1.	Mandatory	Thermometer, Water testing kit, pH meter, Magnetic cleaner, New razor blades, Wiper, Siphoning pipe, Anti-chlorine solution, Acriflavine solution, Antibiotic tablets, Old newspaper, Duster, Air tube, Diffuser stones, Connectors, Regulators, Screw driver, Plier, Electric tester, Fish feed
2.	Essential	Air pump, Heater, Power head, Filter foam, Zeolite rings

Threats and weakness: An aquarium technician should be very careful while servicing an aquarium. The hobbyists will be always blaming them for any mortality of fish. A small negligence in their part if caused fish mortality will affect the future business.

4.2 Fish pedicure spa

These are specialised spa where feet therapy is provided using *Garra rufa,* a fish of Turkish origin. The customers are suggested to put their feet in a water tank stocked with *Garra rufa* in many numbers. This fish is having a unique feeding habit and feeds on the dead skin of a person (Picture-14). So, when someone puts their feet in the tank, the *Garra rufa* will feed on the dead skin of the feet and make it look beautiful and healthy. The fish pedicure spas are very popular in many countries, specifically in Thailand. These spas are becoming popular in India now days generally in shopping malls. The charges of therapy vary from Rs. 100-500/- for half an hour session depending on location of spa. The setting up of a fish pedicure spa specifically in shopping mall or at places of tourist interest could be a promising business activity for our youth.

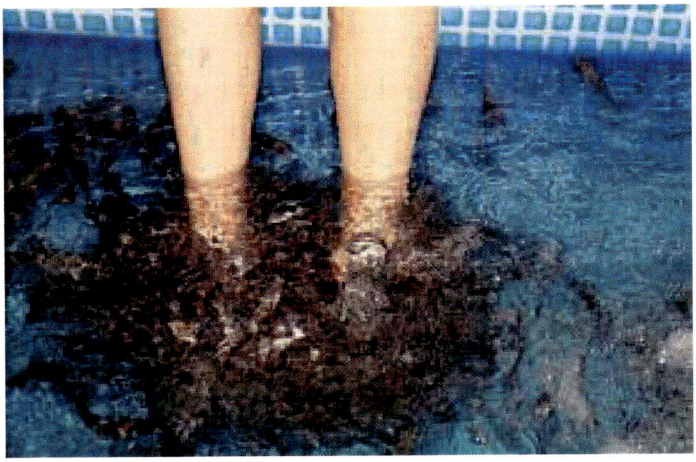

Fig. 7.14: Garras rufa fish pedicure spa

Strengths and opportunities

- Increasing awareness among a common man to look healthy and smart
- The spending capacity of a common man on comfort and luxury is increasing.
- Not a highly technical activity.
- A good unit can be installed at an initial cost of Rs. 1.00 lakh only.

Threats and weaknesses: Most of the fish pedicure spa installed in the country are not hygienic therefore their use is avoided. The *Garra rufa* is not included in the list of fishes permitted for import by GOI. The mortality of fish is also very high.

4.3 Setting up of a Public Aquarium Gallery

The Public Aquarium Galleries (PAG) are centres of entertainment and education to a visitor and a good source of earning to the owner. There are many numbers of very large PAG around the world specifically at tourist's destinations. The "Taraporewala Aquarium" was the first PAG in the country set up in year 1951. It was the only well-known PAG till the end of 20th century. Presently, there are several small setups at several zoological parks and a few PAGs have been established in different parts of the country viz; Ahmedabad, Jammu, Surat, Lukhnow, Bhubaneshwar, Udaipur, Ratnagiri, Ranchi, Port Blair etc. All these PAG's are very small in comparison to the present international standards and inadequate in numbers considering the size of the country and number of important tourist destination. There is no PAG even in a city like Goa which is visited by lakhs of national and international tourists every year. The setting up of few international standard PAGs and several small PAG at important tourists' destinations of the country could be very good business option. A PAG of international standard has been set up in science city Ahmedabad since 2021 which is presently the largest in country (Picture-15).

Fig. 7.15: Public aquarium gallery at science city, Ahmedabad inaugurated on 16th July, 2021

Strength and opportunity

- There are only few PAG in country even at important tourist's destinations.
- There is only one PAG in country at Ahmedabad which is of international standard.
- High foot fall of visitors in PAG specifically at tourists' destinations.

Threats and weaknesses: A traditional set up of PAG may be successful in small cities but not encouraging at important tourist's destinations. The cost of an international standard PAG is very high and will require high engineering and business management skills.

Conclusion

The ornamental fish and aquarium industry of India is following a cyclic relationship of demand and supply. The increasing number of aquarium hobbyists is triggering the setting up of new ornamental fish production units and aquarium shops which in turn adding many more new aquarium hobbyists. The increasing competition among the suppliers and traders has caused lowering of basic aquarium cost. So, the aquarium keeping hobby have become less expensive and affordable for large number of families. The process of cyclic relationship between demand and supply may remain continue for many more years in domestic aquarium industry. The hobby of ornamental fish keeping will become much easier to practice as the availability and choice of products will further increase. The demand of hobbyists is also changing from low and medium quality fish to high and premium quality fish as the hobbyists are becoming much more mature and knowledgeable about fish keeping. The growth trend of domestic aquarium trade in India indicates that the number of aquarium hobbyists in India will be highest to any developed country in next 8-10 years.

References

Jain, A. K., 2009. India needs new generation public aquaria and ocenaria. Fishing Chimes., 29 (2): 39-43.

Jain, A. K. and Jain, A. 2009. Domestic aquarium trade is an avenue of immense opportunities and development potential. Proceedings of National Workshop on "Advances in Aquaculture and Fisheries: Perspectives, Prospects and Challenges" at Pragati Maidan New Delhi., 3rd July, 2009: 93-99

Jain, A. K., 2016. An overview of freshwater ornamental fish production industry in India. In : (Eds: Jain, A. K., Saini, V. P., Kaur, V. I.,), Best management practices for freshwater ornamental fish production. National Fisheries Development Board (Ministry of agriculture and farmers welfare, Government of India), Hyderabad, India. 7-20.

Jain, A. K. 2017 A "Code of conduct" needs to be practiced by all the producers, traders, and hobbyists of ornamental fish for sustainable development of industry and contribute to a skilled, clean, and healthy India. 13-15 May, 2017, Aqua Aquaria, India, Mangalore, Karnataka, The Marine Products Export Development Authority (Ministry of Commerce and Industry, Government of India, New Delhi), Kocchi 217-226.

Silas, E.g., Gopalakrishnan, A., Ramachandran, A., Anna Mercy, T.V., Kripan Sarkar, Pushpangadan, K.R., Anil Kumar, P., Ram Mohan, M.K. and Anikuttan, K.K. (2011). Guidelines for Green Ceertification of Freshwater Ornamental Fish. The Marine Products Export Development Authority, Kochi, India.106 p.

8

Freshwater Pearl Culture A Novel Idea to Boost up Aqua-economy

E.M. Chhandaprajnadarsini[1], Shailesh Saurabh[1] and Sambid Swain[2]

[1]*ICAR-Central Institute of Freshwater Aquaculture, Kausalyaganga, Bhubaneswar-751 002, Odisha, India*
[2]*School of Fisheries, Centurion University of Technology & Management, Paralakhemundi Campus, R. Sitapur, Odisha-761211*

Abstract

Freshwater pearl culture emerges as a promising avenue for boosting the aqua-economy by providing employment and ensuring food security. Traditionally, natural pearl harvesting from various regions depleted resources due to indiscriminate fishing and pollution. To address this, cultured freshwater pearl culture has become a successful aquaculture practice, offering economic benefits, positive environmental impacts, and nutritional advantages. India, with its abundant natural water bodies, holds significant potential for freshwater pearl culture, which could reduce the country's reliance on pearl imports. The freshwater pearl culture involves collecting mussels, pre-operative conditioning, surgical implantation of nuclei, post-operative care, pond culture, and harvesting. Ideal water quality parameters, including pH, alkalinity, hardness, temperature, and dissolved oxygen, contribute to successful pearl farming. The economic viability of freshwater pearl farming is underscored by its profitability, lightweight and non-perishable final products, and support from government agencies (PMSSY) through subsidies. ICAR-CIFA plays a crucial role in disseminating technology and training farmers. The sector's challenges include long-term commitment, mussel survival, nutrition, water quality management, location selection, and skilled manpower for surgical implantation. The prospects for freshwater pearl farming in India are promising, with ongoing efforts to raise awareness and develop the industry sustainably.

Keywords: Mussel, pearl, implantation, water quality, nucleus

Introduction

Culture of aquatic organisms is a significant economic endeavor that benefits millions of people worldwide by generating employment opportunities and ensuring food security. One of the lucrative aquaculture ventures in many countries is the cultivation of freshwater pearls. Prior to the development of cultured pearl farming, pearls were harvested from the Gulf of Mannar, the region between India and Sri Lanka, the Bay of Bengal, the Egyptian coast (Red Sea), and the Persian Gulf (Saudi Arabian coast) (Matlins 2001; Strack 2008), which benefited the nation's economy. However, the natural pearl fisheries have been depleted as a result of indiscriminate fishing, habitat destruction, and pollution. Realizing its economic significance, pearl farming has developed into a successful aquaculture activity to support the nation's aqua-economy through innovation and standardization of farming techniques.

Pearls are cultured in both marine and freshwater environments, and they are particularly created in bivalves with an inner nacreous layer. Many countries like China, Japan, Australia, Indonesia, French Polynesia, Cook Islands, Philippines, India, Sri Lanka, Bangladesh, Myanmar, Thailand, Malaysia and Mexico are major pearl producing countries through farming (Gervis and Sims 1992; Southgate 2007; Southgate et al. 2008 & Saurabh et al. 2021). In addition to the economic benefit like other bivalves these pearl producing mussels has positive environmental impact and also has several nutritional benefits. Freshwater bivalves including these pearl mussels are widely used in aquaculture for aquaculture-effluent treatment of aquaculture farms. The shells of bivalves are the source of two highly versatile chemical compounds, calcium carbonate or calcium oxide which is having wide industrial applications. The shells are used in cosmetics and traditional medicine, handicraft and jewelry. These economic and environmental benefits have catalyzed expansion of the bivalve aquaculture industry in various regions of world. During 2009 to 2018 the average freshwater pear cultured production was 1928 tonne while China continued to dominate world freshwater pearl culture (FAO, 2020). However, the freshwater pearl culture production has declined from 3321 tonne in 2009 to 703 tonne in 2018. This drastic decline may be due to overexploitation, habitat degradation, pollution, lack of proper management and many more and it has threatened long term sustainability of fisheries and this farming sector. It gives an alarming call for all fisheries managers and policy makers to formulate management measures to implement for the betterment of the sector. Though this farming technology has a great potential to increase the aqua economy of the country, growth should be sustainable.

Pearl is one of the "Navaratnam" and it known as "Queen of Jewels". It is considered as a symbol of wealth from days of old. India has a great potential of freshwater pearl culture as it is bestowed with vast area of natural water bodies like ponds, rivers, lakes, reservoirs. However, our country spends so much money on importing pearls from other countries. In recent years, pearl farming is being promoted more keenly by the Fisheries state department as well as central government through NFDB, Hyderabad to boost up the aqua-economy and to create employment opportunity. Compared to the marine pearl production fresh water pearl production require lower capital and operational cost with less resource demanding (Rachman & Maskur 2006, Johnston et al. 2019, 2020). Pearl farming especially designer/image pearl farming is very highly profitable farming practice and it is appropriate for expansion in rural areas with low input and high output cost (Hossain et al. 2004).

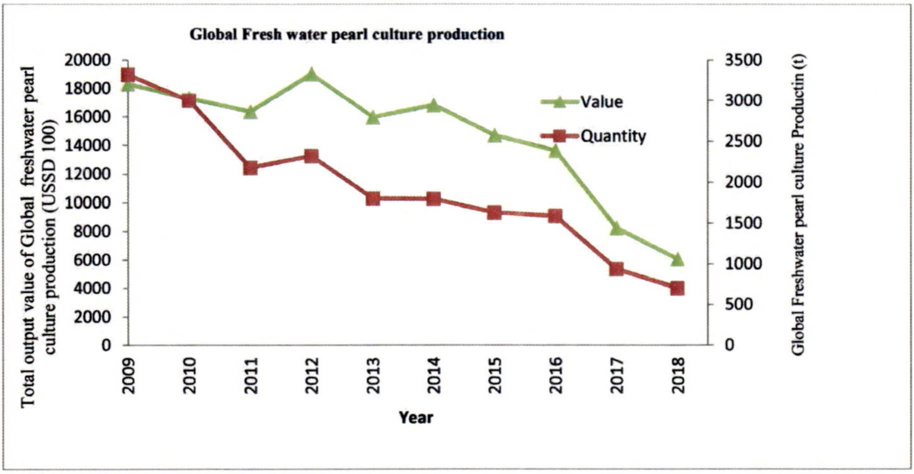

Fig. 8.1: Global freshwater pearl production

Species diversity and Distribution

There are about 52 species of freshwater mussel inhabit in India. Among these, 3 species *i.e* common freshwater mussels, pond mussel *Lamellidens marginalis,* paddy field mussel *L. corrianus* and riverine mussel *Parreysia corrugata* are suitable for pearl culture in India (Janakiram 1989). Generally, *Lamellidens sp.* prefers to inhabit upto 0.5-1 m depth in stagnant to slow flowing ponds and reservoirs. They are mostly found in ponds with soft sediment preferably alluvial soil and the water should be rich with green algal bloom (*Chlorococcum* sp. & *Scenedesmus* sp.) But *P. corrugata* prefers lotic habitat mostly rivers (Janakiram and Radhakrishna 1984). They are widely distributed in many parts of India *viz.* the northeast, western, central, and southern states of India (Thomas 1974). Water body with hard substratum, macrophyte and algal bloom *Euglena* sp. and *Microcystis* sp is not suitable for pearl mussels.

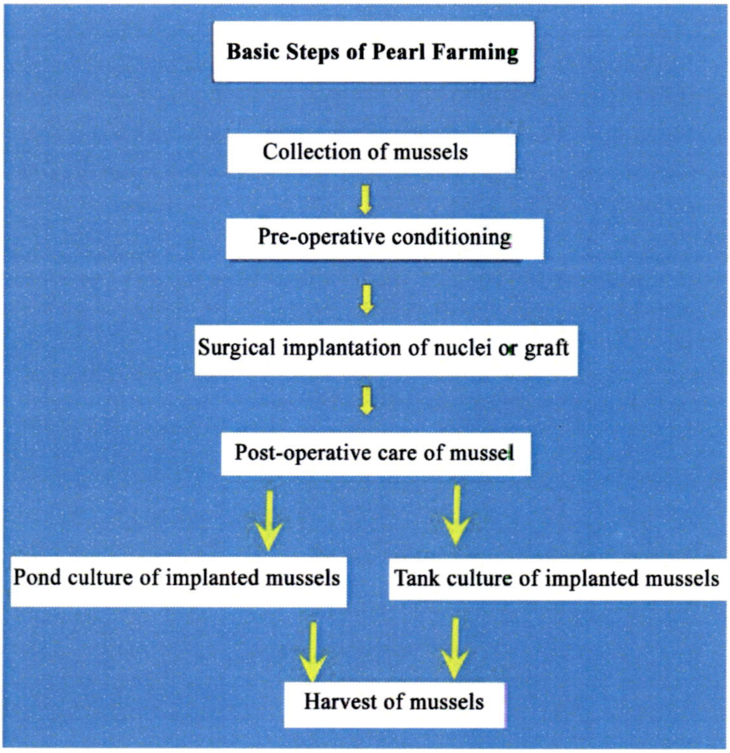

Fig. 8.2: Basic steps in pearl farming

Basic steps of Freshwater Pearl Culture

Collection of mussels

Freshwater pearl mussels *L. marginalis* and *L. corrianus* 8.0-11 cm in shell length and 50-60 g wet weight were hand-picked from natural habitats and brought to laboratory within 2 to 3 hours. For short-distance transport, mussels were brought in moist condition. However, to prevent dehydration and damage during longer shipment, each mussel was individually wrapped in damp cotton linen.

Pre-operative conditioning

Crowding of mussels (at the rate of one mussel/L) for 24 to 48 hours in ferro-cement or FRP containers (200 L capacity) ensured proper relaxation of the adductor muscles, facilitating easy opening of the shell valves for surgical implantation. Studies made at ICAR-Central Institute of Freshwater

Aquaculture (CIFA) have indicated that narcotizing procedures followed for marine oysters with menthol crystals (James 1991) were not applicable for freshwater pearl mussels. It was observed that freshwater mussels remained closed in the containers when menthol crystals were sprinkled over the water surface.

Surgical implantation of nuclei or graft

In India, the surgical procedures described by Janakiram and Tripathi (1992) and Saurabh et al. (2014) are mostly followed. The primary requirements for surgical implantation are specialized surgical kits, nucleus beads of various shapes and sizes, and mussels with nacreous layers. The hard molluscan shell powder, egg shell powder, or acrylic material used to make the nuclear beads. Nucleus preparation is a rather straightforward experience. Utilizing widely available self-curing acrylic repair material, the nuclear beads are generated (powder and solvent). The other kind of nucleus is constructed using glue that is readily accessible on the market and shell or egg shell powder. Different sorts of designer mould are offered on the market to create nuclear beads in a variety of designs.

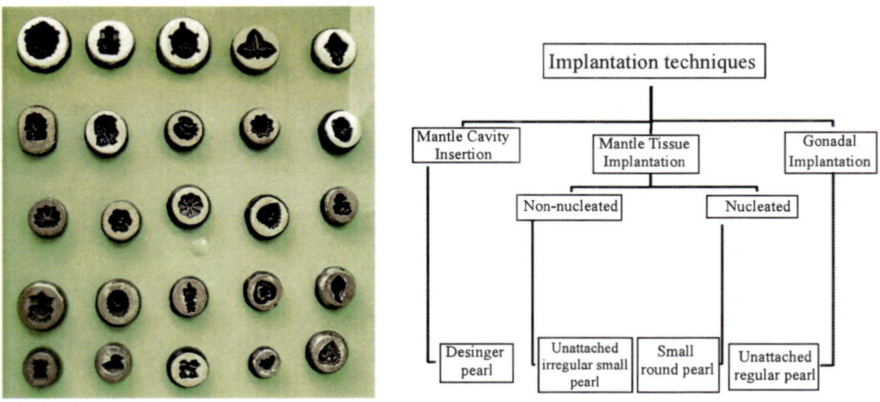

Fig. 8.3: Implantation technique

The mantle cavity insertion method involves inserting a shell bead of a variety of forms into the mantle cavity of pearl mussels. The implanted mussels are then raised for a year in an aerated tank or cultured pond. In the mantle tissue implantation procedure, the donor mantle graft (2mm ×2 mm) which was processed in a hygienic environment is carefully inserted in the pockets of mantle tissue of the recipient live pearl mussels with the help of graft insertion cum lifter needle. Generally, 2 to 4 pockets are made in the posterior lobe of the mantle tissue on both left and right side of recipient mussel. The diameter of the round nucleus, which is likewise implanted alongside a graft, is determined by the thickness of the mantle tissue. The mussels that have been implanted are raised for a year in the pond habitat and or specially designed FRP tank having aeration and water exchange facility. In gonadal implantation the donor mantle graft (2-4 mm×2-4 mm) along with nucleus (3-6 mm dia.) is placed in the middle of the gonad of the recipient live mussel. To reduce the likelihood that the implanted graft and the nuclei would be rejected, the mussels with gonad implants were kept in post-operative care with a broad-spectrum antibiotic supplement for 7 to 10 days. The implanted mussels are reared in the pond environment for a period of 15 to 18 months in order to get better coating of nacreous substances and that in turn will add extra value of the finished products. A gentle flow of water in tank condition and feeding with green algae *Chlorella* will ensure better coating of nacreous layer which will result in a glittering pearl.

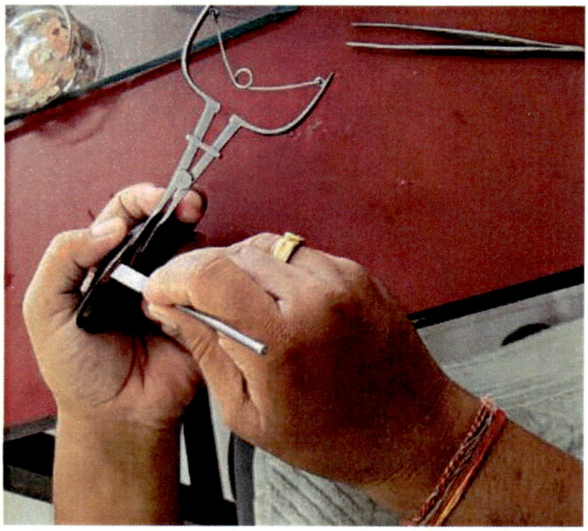

Fig. 8.4: Implantation

Post-operative care of mussel

Following surgery, the mussels are placed in a post-operative clinical setup with adequate illumination and water recharging options. The mussels were tied in the nylon bag (2 mussels per bag) by keeping the ventral side up to reduce the rejection rate of nucleus. The bags are kept in the FRP tank (200 L capacity) at the rate of 50 mussels/tank replenished with aerated aged tap water at a depth of 0.2 m. A provision of aeration facility in the care unit will ensure better health of implanted mussel. This phase typically lasts for 7–10 days in order to heal any wounds that may have happened after surgery. To combat the infection and improve survival during this period, broad-spectrum antibiotics are administered. The implanted mussels are fed with phytoplankton rich water on regular basis.

Pond culture

Implanted mussels can be cultured in a well-managed traditional carp culture pond with 2.5 m depth. The pond with clay soil base and slightly alkaline water are suitable for pearl farming. The bamboo-raft are made in the pond to hang the implanted mussels. The ideal stocking density of the implanted mussels is 25000/Acre. Periodically the implanted mussels are cleaned and condition of the mussels are checked for better management. The commonly preferred algal species by the fresh water mussel are *Chlorococcum, Chlorella, Spirulina.* The ideal water quality parameter for the pearl farming are listed in the following table.

Table 8.1: Ideal Water quality parameter for Freshwater Pearl Farming

Parameter	**Range**
Transparency of water	40-60 cm
pH of the water	7-8
Temperature of culture medium	25-30 ^{0}C
Dissolved oxygen	4-8 ppm
Total Hardness	60 ppm
Calcium in water	20-30 ppm
Total Alkalinity	80 ppm
Ammonia	Less than 0.05 ppm

Harvest

After 12-18 months of culture period implanted mussel were harvested from the culture unit. For the designer pearl, mussels are sacrificed, cleaned and process for value addition. For mantle tissue implantation and gonadal implantation mussels, pearl can be removed from the live animal through surgery.

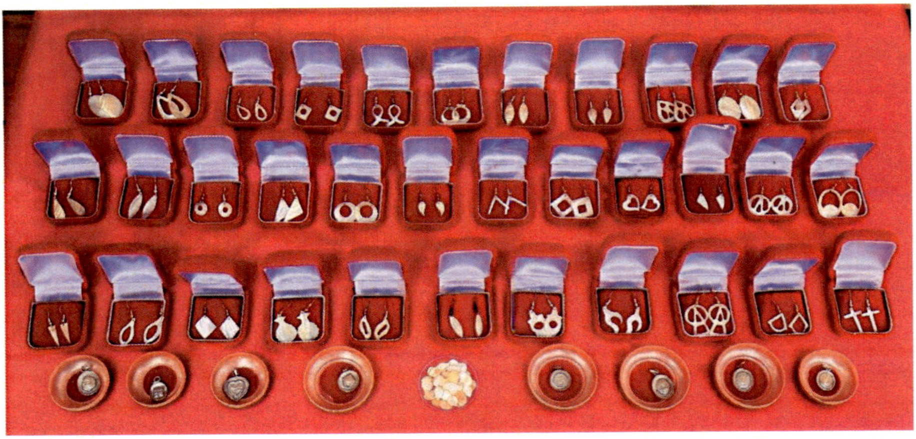

Fig. 8.5:: Value addition of harvested pearl

Economics of Freshwater Pearl Farming

Pearl farming is one of the best profitable aquaculture ventures not only in India but also in all over the world because of the huge demand in the local as well as export market. The best part of farming is the final product is lightweight, portable and non-perishable unlike other aquaculture practices. Also, the pearl farming can be done as a monoculture and along with other farming practices. Farmers can do freshwater pearl farming in a small designed tank having aeration facility as well as in well managed pond. Realizing the economic benefits of the pearl farming, NFDB, GOI and other government agency provides subsidies to the farmers to start pearl farming and aquaculture. In various states, several schemes for establishment of farm and operational cost are being provided by the state fisheries Department. ICAR-CIFA, Bhubaneswar has played a significant role in popularizing freshwater pearl culture technology by disseminating the technology to the interested stakeholders. The scientist of this institute provides hands-on training on cutting edge techniques of freshwater pearl farming in a custom created training programme both in offline and online mode. Few trained pearl farmers of this institute who had begun acquiring the necessary knowledge and abilities for small-scale pearl farming also mentoring and instructing other beginning farmers in order to spread the technology to the remote areas of the country. The product range of pearl farming are very wide such as round pearl, shell attached mabe pearl, designer pearl and baroque pearl. Market value of the pearl depends on size, uniformity in the nacre deposition and lustre. However, designer pearl with good lustre and perfect uniformity in nacre deposition fetches a comparative better market value than other regular round or irregular pearls. Freshwater pearl farming has proven to be one of the lucrative ventures for the industry development in many rural areas in India for the establishment of cottage industry.

In the year 2008, Janakiram gave a cost estimate for a small-scale commercial farm of 0.4 ha area for 2-year culture period. For this venture the input cost was about 6.75 Lakh while the return was 10.5 lakhs. In Rajasthan pearl farmers are earning nearly about Rs. 59,958.728 (without subsidy) and Rs. 1,07,642 (with Subsidy) annually (Yadav & Sharma, 2022). Under ideal management practice, the earning can be increased. A lay out of the model cost return of the pearl farming with best management practices is given in a tabulated format below. Please note that the cost may change on timely basis and regional basis.

Table 8.2: Model Cost Estimates of Freshwater Pearl Culture in Earthen Pond (1 Acre)

Particulars	**Nos**	**Amount (Rs in Lakh)**
Construction cost of pond (2×0.5=1 acre)	2 nos.	3.00
Mussel holding tanks 20 FRP tanks of 250 L capacity @ 3,500/tank) For preoperative and post-operative conditioning	20	0.70
Wet laboratory cum Implantation Chamber (12 x 5 m) with furniture sets (2)	1	1.70
Overhead tank	2 no.	0.50
Surgical Tool kit for implantation (2 sets @ Rs. 20,000/set)	2	0.40
Culture units (PVC) Pipes and floats, Bamboo, ropes)		0.20
Pump for water facility	2	1.50
Air blower (1HP), Air stones and accessories	2	0.25
Electrification and Plumbing		0.50
Sub Total		**8.75**
Participation in Skill development training on Freshwater Pearl farming (Nodal Centre: ICAR-CIFA, Bhubaneswar)		0.10
Mussels Seed (@ Rs. 5/mussel)	25,000	1.25
Designer pearl nuclei (@ Rs 2/nucleus)	50,000	1.00
Water quality testing, Fertilizers, lime, feed etc		0.40
Manpower 1 nos @ 12,000/ p.m for 12 months	1	1.44
Miscellaneous costs viz glue, medicine, mortar and pestle, Transportation , pearl certification etc.		0.50
Post-harvest processing and ornamentation of pearls		1
Sub-Total		5.69

Table 8.3: Cost and Returns in Freshwater Pearl Farming in 1-Acre Pond

Sl.No.	Particulars	Pond (1 Acre)
	Cost	**Amount (Rs in Lakh)**
1.	Capital cost	8.75
2.	Recurring/operational cost	5.69
3.	Total project cost (A)	14.44
4.	Depreciation @ 15% on Capital Cost (B)	1.31
5.	Interest @ 12% on Total Project Cost (C)	1.73
6.	Grand total (TPC + Depreciation + Interest)	17.48
7.	Gross income (60 % Survivability) Farm gate price of each designer pearl @ Rs 180 (A)	21.60
8.	Net profit after the end of the year Gross income-Grand total	4.12

Following points needs to be considered before planning to start freshwater pearl farming on a commercial basis

- The pearl farming is a long-term process and it needs more time at least a year and diligent work in addition to the initial financial commitment.
- Survival of nucleated mussels is a major impediment of the farming. According to study done by various researchers around the world, less than 70% of freshwater mussels survive surgical implantation under normal conditions, and of those, 30–40% are likely to reject the implanted nucleus, while 20% will produce marketable pearls and only 5% will produce top-quality gemstones (Fassler, 1992; Norton et al. 1996).
- Just like other farming techniques, freshwater pearl culture requires adequate nutrition and water quality management. Since the quality and quantity of the pearl depend on proper management, particularly with regard to water quality and a balanced diet, this is quite important. Maintaining pH and ammonia levels is the key challenge in pearl farming. The pH level of the water should be kept at 7-8 and the ammonia level should be zero (Bhargava et al., 2020). In addition, calcium levels in water should be one of the essential elements to ensure uniform nacre deposition on finished goods.
- For farming purposes, one ought to pick a good location free of predators and pollutions.
- Access to a reliable source of freshwater mussels for culturing.
- Surgical implantation requires skilled manpower and is a skill-based method. Without receiving instruction from an accredited institution

like ICAR-CIFA, some farmers have been observed attempting to do surgical implantation on their own, which leads to significant mussel mortality and financial losses.

- Create a successful marketing strategy to sell the pearls, as this is a significant issue in the sector.

Prospects of Freshwater Pearl Farming in India

In India, pearl farming is a growing economic potential. Several states, including Odisha, West Bengal, Bihar, Assam, Chhattisgarh, Maharashtra, Gujarat, and Kerala, contributed to the development of this farming technology. It is still in the early stages of development in certain states. Even though this technology is quite profitable, compared to other sectors of aquaculture, farmers have not embraced it to the same extent. The lack of a structured sector, being unaware that mussel seed stock is available in their localities, inconsistent breeding practices, improper management ofbrood stock, mortality of implanted mussels, and many other factors may all be contributing factors. However, ICAR-CIFA is playing a crucial role in spreading this technology to the impoverished farmers and concentrating on high end research on the aforementioned concerns in order to develop the industry in a sustainable manner. In an effort to develop the country's R & D facilities, every year ICAR-CIFA administers a range of training programme for farmers, entrepreneurs, assistant professor, KVK Official, State Fisheries Department Official and other stakeholders. The production process of quality pearls in a shorter amount of time is being sped up through research in this institute. It is now necessary to raise public awareness of freshwater pearl farming and its economically viability in order to strengthen the aqua economy of the country.

References

Bhargava, C., Singh, S., & Ambulkar, P. L. (2020). Case study on freshwater pearl cultivation at tribal district Chhindwara, Madhya Pradesh, India. *Science Archives,* 1(3), 117-122.

FAO. Fisheries and Aquaculture Statistics. Food and Agriculture Organization of the United Nations, Rome; 2020. 248p

Fassler, C. R. (1992). [Pearl culture, a new opportunity arise to entrepreneurs beyond Japan]. [Italian]. Selezione Veterinaria.

Gervis, M. H. & Sims, N. A. (1992). The biology and culture of pearl oysters (*Bivalvia pteriidae*) (Vol. 21). WorldFish.

Hossain, M. A., Sultana, N., Azimuddin, K., Hussain, M. G., & Mazid, M. A. (2004). Selection of freshwater pearl mussel species for mantle transplantation in Bangladesh. *Bangladesh Journal of Fisheries Research,* 8(2), 2004, 113-116.

James, P. S. B. R. 1991. Training Manual on Pearl Oyster Farming and Pearl Culture in India. Training Manual No. 8, Tuticorin, India: Central Marine Fisheries Research Institute.

Janaki Ram, K., & Radhakrishna, Y. (1984). The distribution of freshwater mollusca in Guntur District (India) with a description of *Scaphula nagarjunai* sp. n.(Arcidae). *Hydrobiologia*, 119(1), 49-55.

Janakiram K (1989). Studies on cultured pearl production from freshwater mussels. *Current Science*, 58 : 474-476

Janakiram, K. (2008). Freshwater Pearl Culture Technology Development in India. *J. Applied Aquaculture,* 13(3-4): 341-349.

Janakiram, K., & S. D. Tripathi, S. D. (1992). A Manual on Fresh-water Pearl Culture. Manual Series 1, Central Institute of Freshwater Aquaculture, Bhubaneswar, India

Johnston, W., Gordon, S. E., Wingfield, M., Hine, D., & Southgate, P. C. (2020). Economic feasibility of small-scale mabé pearl production in Tonga using the winged pearl oyster, Pteria penguin. Aquaculture Reports, 17, 100347.

Johnston, W., Hine, D., & Southgate, P. C. (2019). Overview of the development and modern landscape of marine pearl culture in the South Pacific. *Journal of Shellfish Research*, 38(3), 499-518.

Matlins, A. L. (2001). The pearl book: the definitive buying guide: how to select, buy, care for & enjoy pearls. Gemstone Press.

Norton, J. H., Dashorst, M., Lansky, T. M., & Mayer, R. J. (1996). An evaluation of some relaxants for use with pearl oysters. *Aquaculture*, 144(1-3), 39-52.

Rachman, B. & Maskur, T.W. (2006). Effect of various depth against implantation process and nucleus coating pearls in oysters Margaritifera sp. on controlled pool. Impasja, 2, 86–95.

Rachman, B., Winanto, T., Maskur, M., & Sukmajaya, Y. (2017). Pengaruh Kedalaman Terhadap Proses Pelapisan Inti Bulat Pada Kerang Air Tawar (*Anodonta woodiana*). *Jurnal Biologi Indonesia*, 6(1).

Saurabh, S., Pradhan, S., & Suman, S. (2021). Recent Trends in Freshwater Pearl Farming in India. In Update on Malacology. IntechOpen.

Saurabh, S., Mohanty, U. L., Mohanty, J. & Jayasankar, P. (2014). Pearl Culture Technology in Freshwater Environment. In: Aquaculture and Fisheries Environment (S. K. Gupta and P. K. Bharti, eds.). Discovery Publishing House PVT. LTD. New Delhi-110002, pp. 51-78.

Southgate, P. C., & Southgate, P. (2007). Marine Pearl Industry. *Pearl Oyster Health Management: a manual*, (503), 7.

Southgate, Paul C., Elisabeth Strack, Anthony Hart, Katsuhiko T. Wada, Mario Monteforte, Micheline Cariño, Sandra Langy, Cedrik Lo, Hector Acosta-Salmon, & Aimin Wang. (2008). "Exploitation and culture of major commercial species." *The Pearl Oyster* 303: 355.

Strack, E. (2008). (2008) Introduction. In: Southgate PC, Lucas JS (eds) The pearl oyster. Elsevier Press, Oxford, pp 1–35

Thomas, E.I. (1974). The bionomics, anatomy and development of the freshwater mussel, Lamellidens marginalis (Lamarck). Annals of Zoology Quarterly (Agra, India), 10: 71–169.

Yadav, R. & Sharma, A. (2022). Pearl Farming: An Economically Viable Entrepreneurship Development Opportunity in Rajasthan Entrepreneurship in Livestock and Fisheries pp. 245-255

9

Entrepreneurship Opportunities in Freshwater Ornamental Fishery in India

Archana Sinha

Head and Principal Scientist, ICAR-CIFRI, Barrackpore, Kolkata 700120

Abstract

Ornamental fish are very attractive, colourful, small and easily housed in confined water. These fishes are being kept in glass aquarium to observe the beauty and their movements for happiness and mental peace. Keeping pet in houses is a hobby of mankind. It is well known that ornamental fish keeping is the second most expensive hobby after photography. The demand of varieties of ornamental fish and other accessories to maintain an aquarium has attracted the entrepreneurs to accept the breeding and culture of ornamental fish as a challenging business. Starting from selling ornamental fishes there are other opportunities to sell glass bowl, aquarium, aquarium accessories, decorative toys, aquatic plants, sand, pebbles, live and artificial feed for the aquarium fishes. This sector provides employment to the youth especially women. Keeping the potential of this business in livelihood security in mind, Government of India has extended financial assistance to the entrepreneurs.

Keywords: Freshwater ornamental fishery, Ornamental fish, Ornamental fish trade

Introduction

Ornamental fishes have been recognized as the important avenue for the income and employment generation in the country. The economic activities of ornamental fish culture involve a complete chain of the entrepreneurs and enterprises i.e. fish collectors from natural water bodies, local traders, exporters, manufacturers of the aquarium, fish breeders, ornamental fish farmers, vendors etc. With the increase in income and urbanization, the habit of keeping ornamental fishes has been increasing at rapid pace. However, there are no consitenet markets for Indian ornamental fish and a regular fluctuation with respect to destinations is reported since the geographical spread of the markets has exhibited an ever changing hue. The top markets of ornamental fish from India are Singapore, USA, Hong Kong, Malaysia and Japan, which jointly accounted for about 70% of total India ornamental fish export. Amongst them Singapore alone accounts for almost 43% of total export in ornamental fishes followed by Japan (14%) and Malaysia (10%). USA and Hong Kong both accounted for 7.5% each. However, contribution of India in world market is negligible. In India, the main objectives of ornamental fisheries are employment generation, women empowerment and earning foreign exchange. Keeping the imense potential of ornamental fish trade in mind the sector is identified as one of the thrust areas for development to augment exports in the country. To prosper any trade there are three pre-requisites called sustainability, quality and quantity.Potential natural resources of indigenous ornamental fish of high value in the country has greatly contributed in development of ornamental fish industry.It provides enhanced cash income and thus provides better living standards for those involved in the trade. About 400 species of marine ornamental fish and 375 freshwater ornamental varities are available in varoius parts of our country. The domestic trade of ornamental fish is mainly dominated by production and marketing of low value ornamental fish like Gold fish, Koi carp, Sharks, Gourami, Angels, Barbs, Live bearers etc. The retail prices for these fishes vary around INR 15-50 each. The sustainable growth in this segment is possible through captive breeding to ensure a constant supply of hatchery-bred young ones to the global markets round the year. Breeding and culture of ornamental fish can be a promising alternative livelihood for women and unemplyed youth. The key factors for success in commercialization of ornamental fish farming are cluster/group approach in farming with wide species variability, production scale, and quality. The price competitiveness is limiting the middlemen to possible extent in the marketing and overcoming the supply or delivery limitations. Recognizing the potential and possibilities, Government of India has envisaged a program to unlock the country's ornamental fisheries sector through implementing different activities in this sector. Implementation of Ornamental Fisheries on pilot basis, in its scope and reach, focuses mainly on creating an enabling environment for a sustainable and holistic development and management of ornamental fisheries for the socio-economic benefit & development of ornamental fisheries in India. Thrust areas have been identified for enhancing ornamental fish production through farming and conservation of natural resources through habitat restoration and creating mass awareness amongst the stakeholders.

Trades associated with Ornamental Fishery

Ornamental fishery is getting popularity as a hobby among families of middle and high income groups of the country. The hobby has attracted entreperuners to earn cash income by providig the services to the hobbyists. The Public Aquarium is one of the major attarction for number of individuals especially children and aged. Now, a considerable number of families are keeping an aquarium in their houses as a hooby. The important sectors of the emerging bussiness of ornamental fish are enlistes as-

- Breeding and culture of ornamental fish
- Aquaria designing and fabrication
- Aquatic plants for aquarium
- Decorative toys and accessories
- Formulated and live feed
- Export of ornamental fish
- Aquarium Service providers

Fig. 9.1: A weekly pet market in Kolkata (WB)

Breeding and Culture of Ornamental Fish

The culture of colourful attractive fish in glass aquarium for aesthetic use is called ornamental fish culture. There are more than 250 varieties have been recorded from the domestic market which is dominated by the exotic fish in the Indian market. More than 200 freshwater fish species are recorded which are being reared and bred by ornamental fish farmers including hobbyists in captivity in different parts of our country. The demand for various types of ornamental fishes within India is estimated worth at Rs.550.00 crore in a year with 20% increase in demand every year.Recent trend in the production and demand shares that during next five years domestic trade will become two fold of the present. Ornamental fishes are mainly categorized into two groups namely, egg layers (oviparous) and live bearers (ovo-viviparous). Based on the breeding behaviour of fish, the breeding technology and the requirements of infrastructure differ from species to species.

Most of the ornamental fish species are egg layers and normally exhibit external fertilization. Based on the nature of laying eggs they are categorized as egg scatter which lay non-adhesive eggs or adhesive eggs to the substrate viz. plants or allow them to float on the surface. The egg-scatters either spawn in pairs or in groups. There is no parental care given in fact they eat their own eggs. The number of eggs is more in comparision to other species. Goldfish (*Carrasius auratus*) and tetras are the best example. Fishes categorized as egg depositor either laid eggs on a substrate, like stone or plant leaf or fine leafed plants like java moss. Egg depositors are of two types- one that care for their eggs and another that do not care. The example of egg depositors that care their eggs are cichlids and some catfishes. Cyprinids and some other catfishes come under the egg-depositors group that does not care for their young ones. These species lay their eggs on a surface and the eggs are abandoned. These species usually do not eat their eggs. *eg*. Angel (*Pterophyllum scalare*), Discuss (*Symphysodon discuss*) and some catfishes. Fishes of the group categorized under egg buries are usually lives in the seasonal waters that dry up for some time of the year. The majority of eggs burrier lay their eggs in mud. The maturity time for parents is very short and they lay eggs before dying when the water dries up. These eggs remain in a dormant stage until rain stimulate hatching. *eg*. Annual killifish.Mouth-brooder fish carry their eggs or larvae in their mouth. Mouth brooders can be grouped into ovophiles and larvophiles. Ovophiles are egg loving mouth brooder. They lay their eggs in pits and then female sucked up into her mouth. The small number of large eggs hatch in the mother's mouth and the fry remain there for some time. Many cichlids and some labyrinth fish are the example of ovophiles mouth-brooders. In contrast, Larvophile or larvae loving mouth-brooder fish lay their eggs on a substrate and guard them until the eggs hatch.

After hatching the female picks up the fry and keeps them in her mouth. When the fry can feed for themselves, they are released *eg*. Cichlids. An important group of egg layers is the fish that build different types of nest for their eggs. The nest ranges from a simple pit dug into the gravel to the elaborate bubble nest formed with saliva coated bubbles *eg*. gouramis, anabantids and some catfishes.

Among live bearer fish, which give birth of their young ones without laying eggs, the important ones are Guppy (*Poecilia reticulata*), Sword tail (*Xiphophorus helleri*), Platy (*Xiphophorus maculatus*) and Mollies (*Poecilia reticulata*). Apart from the common Live-bearer and egg layer varieties, breeders prefer to breed highly specialized varieties such as Oscar, Flower Horn, Tetras, Discus and Cichlids.

Black angel

Goldfish

Flower horn

Guppy

Tiger barb

Blue danio

Rosy barb

Pearl gourami

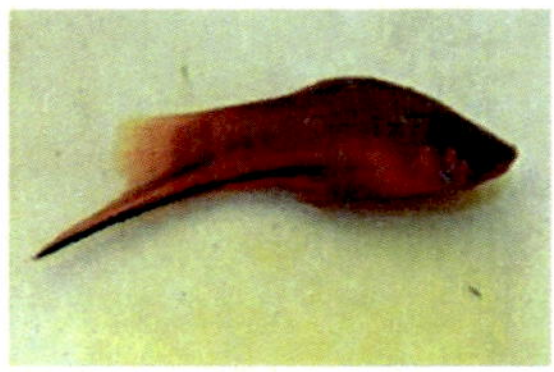
Sword tail

Blue gourami

Siamese fighter

Freshwater shark

Devil catfish

Discuss

Loach

Fig. 9.2: Fresh water ornamental fishes

Aquaria Designing and Fabrication

These lovely fishes are usually kept in aquarium which is a glass tank decorated with special materials to display colourful fishes in natural abode by introducing aquatic plants, rocks, gravel, artificial decorative etc. To maintain the environmental parameters equipments are being used to control aeration, movement of water, temperature, suspended organic matter, illumination etc. Feeding fish in tanks with natural and artificial feed is mandatory. Different types of aquarium, ornamental plants, decorative items etc. are available in the market. It is also possible to construct an aquarium as per the requirements. The simplest tank to keep the ornamental fishes is a fish bowl. It is made up of all-glass having various capacities generally below 5 liters. It serves to accommodate one or a few fishes.

An aquarium tank is the common container used for keeping ornamental fishes everywhere. It is made of glass both from sides and the bottom, fitted into a metal frame of either angle iron or aluminium by means of bitumen. Now a day's all-glass tank are very common due to its varying advantages. Such aquarium tanks are available in many standard sizes. Large tanks are always beneficial to rear big size fishes due to its large water surface for exchange of gases besides adding beauty to the aquarium.

Construction and fabrication of aquaria is an art and it can be taken as a skilled business by entrepreneurs. The design like small bowl, all glass in rectangular shape, moulded ones in different designs, fixed in tables or wall of the houses as interior are in demand. The price of aquarium starts from a few hundred rupees to more than fifty thousands in India. However, this opportunity is not well explored.

Fig. 9.3: Different types of aquariua

Ornamental Plants for Aquarium

Ornamental plants are planted in aquarium not only for beauty but also for well being of the fishes. In addition to decorative value; these plants provide shelter and a sense of security to the fish. They also absorb the injurious gas (Carbon dioxide) given off by the fish in breathing and produce oxygen, which can be utilized by the fish during daytime. Although artificial plastic plants are now available and look extremely realistic, the provision of living plants in the aquarium will undoubtedly help to maintain the biological balance of the environment. A wide variety of plants available for keeping in the aquariums are Vallisneria, Cryptocome, Echinodorus, Ludwigia, Elodea, Hydrophyla, Cabomba, Myriophyllum, etc. The soft-leafed species make good food for herbivorous fishes while the firmer stout-leafed plants offer favorable spawning and is either as spawning sites themselves or as nest building materials. The rearing and growing these aquatic plants in situ and market them is a untoched high return business. Culture aquatic plant in house is a very enjoyable and rewarding occupation. Most established ornamental plant culture businesses started as a small one, and developed into full-time enterprises as the business grow fast. Initially there are a couple of options to consider in developing a market plan for selling the cultured aquatic plants. Since entrepreneurs will start it on a small scale, they can try to develop a few local retail outlets such as garden centers or hardware stores which are selling pond hardware but could use some plants to fill out their product line.

Small scale domestic market

Raised raceways for plants

Plants for sell

Amazon

Aquatic grass

Vallesneria

Cabomba

Hydrilla

Hygrophila

Fig. 9.3: Aquatic plants for aquarium setting

Decorative Toys and Accessories

A big market of aquarium accessories and decorative items exits for aquarium lovers. Starting from rocks, pebbles, toys, aerators, posters, siphon it consists of aerators, thermostat, air stones etc. There are rocks of different design and colours that provide a hiding place to the fishes besides giving beauty to the aquarium. The preferred rocks must be inert like granite, flint, slate, sandstone and well-washed coal. Cement, concrete, chalk or marble must be avoided unless not specially treated. Seashells and corals generally get dissolve in the freshwater aquarium therefore; it should not be used in freshwater aquarium.

Natural rocks can be collected for the aquarium or can be brought from the pet shops. One of the inputs for aquarium decoration is wood in different forms which can also be decorative in the aquarium and it should be used only after performing all the required cleaning procedures. Apart from rocks and wood a number of decorative toys are also available for according attractive look to the aquarium. They include plastic bubblers in the guise of mermaid, underwater diver, oyster shells, angler, human skull, tortoise, frog etc.

Artificial plants

Aquarium accessories

Aquarium toys

Domestic market

Pebbles

Lighting in aquarium

Fig. 9.4: Decorative items for aquarium

Illumination makes the aquarium lively. Proper illumination is important for healthy functioning of a home aquarium. It is very necessary for plant growth and fish activity in aquarium besides viewing the fishes and enjoying the aquarium in habitats. It is observed that too much light generates more algae and too little light retards the growth of plants. Artificial and controlled lighting is desirable for aquarium. A 40-Watt fluorescent lighting for 10 to 12 hours per day is ideal for the normal house aquarium. Now a day's specially designed fluorescent tube lights are available for giving attractive illumination and displaying fishes to their best effect. It attracts entrepreneurs to adopt the business for their cash earning by selling unique colors and styles of aquarium lights.

Surprisingly, there are very few entrepreneurs who are preparing medicines and chemicals for maintaining the fish health. As the ornamental fishes are being kept in captive conditions in small aquarium, chances to encounter diseases is very high. There is demand of chemicals and medicines for the prophylactic treatment and disease control of the ornamental fish by the hobbyists.

Artificial and Natural Feed for Ornamental Fish

Keeping fish healthy in an aquarium is a great challenging task for fish lovers. They have to feed their pet as per their requirement. There are different types of feed available in the market under different trade names. Fish flakes are the most popular fish food in this trade. Slow sinking feed that floats initially on the surface, then slowly settle down and dissolves all over the aquarium, giving the opportunity to eat to all the individuals that prefer lower areas. Pelleted feed are ideal for the species feed in mid-water and fish feeding at the bottom of the aquarium. They are available in different sizes and in regular or slow sinking versions. Their have smooth surface and closed pores, which enables slow water absorption, high hydro stability and limited release of components into the water. Feed tablets for fish have great advantage because they can be glued to the aquarium wall or placed at the bottom. They can be used to feed bottom dwelling fish, crustaceans, as well as mid-water feeders. Fish sticks are floating food for medium and large species of fish that collect it from the water surface. Fish wafers quickly fall to the bottom, and ideal food for crustaceans and fish feeding in the bottom zone. This food absorbs water slowly and thus retains its shape for a long time, which in turn makes it an ideal food for fish whose mouth opening is adapted to scrap the food from different surfaces. Fish chips are shaped like a flat disc and are distinguished by their high hydrostability. Although they are mainly intended for larger species, smaller specimens also have no problem with eating them – they simply bite them off piece by piece. Freeze-dried fish food is a type of food that is somewhere in the middle between natural and artificial ingredients.

It's made form crustacean larvae, insects and zooplankton, free of water. The freeze-drying process consists of freezing these creatures and placing them in a vacuum, which causes the water to evaporate, so that the food retains its nutritional properties, taste and smell. However, natural foods are considered to be the best solution for aquarium fish. When it comes to predatory fish, it means living organisms, and herbivorous aquarium inhabitants are sure to enjoy algae, vegetables or aquatic plants.

There is a potential and demand of low cost feed to rear the ornamental fish. The cost as well as the quality of the feed is concern of the hobbyists. Therefore, there is ample scope to start a fish feed making industry. In India, there is negligible number of feed industry especially for ornamental fish. However, there are different types and makes of artificial feeds available in the market for feeding of ornamental fish. Few examples are: Tokyu baby pellet, Kijaro Basic, Ocean free special fish food, freeze dried tubifex worm, Aquarion, *Spirulina* flake, Freeze dried bloodworms, frozen and flaked brine shrimp, freeze dried krill etc. Recently Growel Feeds Pvt Ltd manufactured commercial feed named *Intan* to satisfy the needs of hobbyists, reducing dependence on imported feeds in the Indian market.

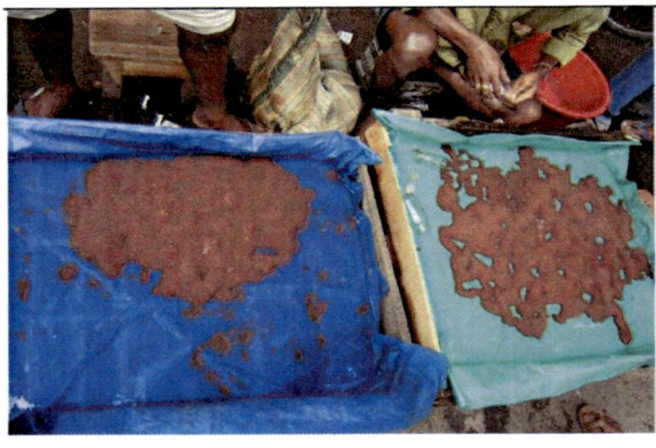

Tubifex worms

Daphnia

Artificial feed of different brands

Feeding a gold fish

Fig. 5: Artificial and natural ornamental fish feed

Export of Ornamental Fish

Ornamental fish trade is practically an unorganized business sector in India although it has considerable scope for export requires immediate attention to conform international standards for maintaining purity in character rather germplasm of fish stock. Existing trend of export is almost dependent on natural source for collection of indigenous species. Such sole and continuous dependence on the natural stock may pose a serious threat rather depletion in stock resulting faunal imbalance. By now the domestic market of such fishes has become already predominant with unknown breed, which cause indiscriminate breeding and no standard breeding technology, is followed. Import of non-certified brood stocks and seed as well is another setback to the industry. India lags behind in creating new varieties of such fishes, a priced activity fully stands through scientific approach as found in the South East Asian countries. Such unscientific approach is in the grip of a few traders who, in fact, create obstacle in allowing scientific farming since they are the root of substandard produce.

In spite of having immense natural ornamental fish resources and technology for breeding and rearing them, not much of headway has been in the country in the matter of export of ornamental fishes to foreign countries. So as to move ahead in these endeavours, the farmers and exporters have to be brought together, for the purpose of integrating the production and export activities in a manner that would be mutually beneficial. The establishment of such a relationship would push up the level of exports of ornamental fishes from the country, particularly to USA, Europe and Japan. It has been reported that 8% of the estimated 86 million houses in USA keep aquaria in their homes, 14% of the estimated 21 million houses in Great Britain, 4% of homes in Belgium and Holland and 5% of German and 20% of Dutch houses keep fish. China, South Africa and several other countries too have the hobby of ornamental fish keeping. In view of the huge demand for export of ornamental, it is possible to undertake mass production of ornamental fish by farmers, to be made available to exporters. In fact producers can become exporters so as to have the advantage of earning foreign exchange themselves. However, it is found that the women cooperative societies, which are breeding and rearing the ornamental fishes, are not getting justice in the market. They are not getting their reasonable share from exporters. It is due to lack of knowledge and communication problem. They are not aware of the export market and the outlets from where they can send the fishes directly. It is important to sensitized them and make aware about the marketing system.

Service Providers for Aquarium Maintenance

The industry is facing the challenges of non-availability of skilled manpower, inadequate training facilities for breeding, farming and raising high valued species, labs for health management and limited exposure to quality standards in international markets. There is huge demand of service provider in aquarium keeping and maintenance especially in metro and major cities. An efficient service provider can attend 3-4 such calls and can earn Rs. 1000-2000 per day. However, there is scarcity of such providers. Government of India has initiated Skill Development Programme and providing certificate after completion of the training as technically Skilled Ornamental fish technician. This degree is valuable and they can provide technical services to individual and public aquaria. A skilled person having NSDC certification can earn without any investment.

The Scheme for Promoting Export Oriented Ornamental Fish Production

Government of India has identified ornamental fish sector as one of the thrust areas for generating employment opportunities and augmenting our foreign exchange earnings. Govt. of India has sanctioned a scheme for the development of export oriented ornamental fish industry in India. As per the MPEDA scheme the beneficiaries will be selected after comprehensive evaluation. The beneficiaries will be given a financial assistance up to 7.5 lakhs (50% of investment) on completing the construction of farm as per the guidelines of the scheme. For the effective implementation, the beneficiaries will be graded in to 3 categories based on the level of investment and details of the grading is as follows - Grade-1 units are micro scale units set up in a cluster comprising a minimum of 10 units. The maximum assistance per unit in the cluster of 10 will be Rs. 75000/- or 50% of the capital cost whichever is less. Grade-2 units are small-scale ornamental fish breeding units, which can be set up as individual units. The maximum assistance will be Rs. 2 lakhs or 50% of the capital cost whichever is less. Grde-3 units are larger individual multi-species ornamental fish breeding units breeding high value ornamental fishes. The maximum assistance will be Rs. 7.5 lakhs or 50% of the capital cost whichever is less. The selected beneficiaries will be trained on various aspects of ornamental fish breeding and successful completion of training will be one of the criteria for providing the financial assistance.

Among the multiple schemes promoted by the National Agriculture Bank for Rural Development (NABARD), ornamental fish rearing which has huge export potential is being encouraged under its fisheries division. The project can be taken up either as a community managed programme by the self help groups or by fish farmers and by individual unemployed youth. The project costing about Rs.7 lakh will be refinanced by NABARD apart from subsidies offered by Marine Export Development Authority (MPEDA).

Considering the potential, National Fisheries Development Board (NFDB) is popularizing ornamental fish keeping by providing financial assistance up to 40% of the approved unit cost (55% to North-Eastern States). National Fisheries Development Board (NFDB) has two components of assistance

i. Setting up of Back-yard unit

ii. Large scale unit

With the introduction of "Blue Revolution" from the year 2016-17, the Department of Animal Husbandry, Dairying and Fisheries , has envisaged a pilot project on ornamental fisheries, where in the action plan for implementation has been outlined to holistic development of this sector. NFDB also sponsors "Hands on training in ornamental fish breeding, larval rearing, fish diseases and bio-security aspects of ornamental fish trade and culture"

With a view to augment domestic demand and boost exports, Pradhan Mantri Matsya Sampada Yojana (PMMSY) support ornamental fish cultivation by establishing production units, introducing commercially important exotic species, importing breeding technology, and extending technological marketing and logistical support to entrepreneurs.

Conclusion

Ornamental fish trade is an lucrative business in India. Though it has started late in our country in comparison to other ornamental fish trade countries, now it is gaining popularity in Indians too. The demand of ornamental fish as well as its ancillary is increasing day by day. Studies observed possibilities of ornamental fish trade in India due to the availability of major resources that can be properly utilized for earning more of foreign exchange to upgrade livelihood of ornamental fish farmers and entrepreneurs. The trade has a sizeable potential for employment too. One such activity-backyard ornamental fish breeding and culture, has been identified and proved to offer immense scope for improving the livelihood of rural community.This may be explored by the way of training imparted and attract more and more manpower in this trade. Strengthening and expanding research and extension in this field is very essential for refining and expanding technology basis of this line of fishery industry.

References

De S S, Ramachandran A. 2011. Marketing Channels in Ornamental Fish Trade in West Bengal.' Fishery Technology 48(2): 163-170.

Dey V K. 2016. The Global Trade in Ornamental Fish.' INFOFISH International 4: 52-55.

Ghosh A, Mahapatra B K, Datta N C. 2003.Hatibagan Haat, Kolkata: The Largest Wholesale Ornamental Fish Market of Eastern India'. Fishing Chimes 23, No. 1: 166-168.

Ghosh A, Mahapatra B K, Datta N C. 2003.Ornamental fish farming – Successful small aqua business in India. Aquaculture Asia 8, No. 3: 14-16.

Ghosh I. 2016. Aquariculture and Sustainable Rural Development: Studies from West Bengal, India. International Journal of Innovative Research & Development 5(6): 375-379.

India Today. 'MPEDA launches scheme for ornamental fish breeding units.' India Today, Kochi, May 15, 2017. Online at https://www.indiatoday.in/pti-feed/story/mpeda-launches-scheme-for-ornamental-fish-breeding-units-926470-2017-05-15(viewed on January, 2018).

Mandal B, Dutta C, Laha U K. 2012. Freshwater Ornamental Fish farming as a tool for Socio-economic development of backward women Self Help Groups (SHG's) to restore livelihood security. International Journal of Advanced Biological Research 2, No. 1 (2012): 95-98.

Pailan G H, Sinha A, Borkar A. 2012. Rose petal meal as a natural carotenoid source for pigmentation and growth of dwarf gourami Colisa lalia. Ani. Nutr. Feed Tech., 12(2): 199-207.

Pailan G H, Sinha A, Kumar M. 2012. Rose petal meal as a natural carotenoid source for pigmentation and growth of rosy barb (Puntius conchonius). Indian J. Anim. Nutr.29 (3): 291-296

Pal S. 2015. 'A study on Diversity of Indigenous Ornamental Fish Species available in Howrah District of West Bengal, India.' International Journal of Applied and Pure Science and Agriculture. 1, No. 8: 137-148.

Sinha A, Asimi O A. 2007. China rose (Hibiscus rosasinensis) petals: a potent natural carotenoid source for goldfish (Carassius auratus L.). Aquaculture Research, 18, 1123-1128.

Sinha A, Prabhakar S K, Singh K K. 2012. Status of Indigenous ornamental fishes in markets of Kolkata. J. the Inland Fish. Assoc. 44(1):102-106.

Sinha A. 2005. Ornamental fish biodiversity of India: The present scenario. In Aquatic Biodiversity in India. (Editors: D.R.Khana, A.K.Chopra, G.Prasad). Daya Publishing House, New Delhi (ISBN: 81-7035-367-X): 38-60.

Sinha A. 2020. Ornamental fish trade in India. Smart Agripost*Fisheries 6(1): 6-10.

The Marine Products Export Development Authority, 2004. A Model Bankable Scheme for Setting Up of Export Oriented Ornamental Fish Breeding Units. 21p.

10

Status and Opportunities of Mariculture in India

Imelda Joseph

ICAR- Central Marine Fisheries Research Institute (CMFRI)
Box No.1603, Ernakulam North P.O., Kochi- 682 018, Kerala

Abstract

Aquaculture is considered as the fastest-growing food production sector in the world and currently it supplies over half of all seafood consumed globally. Mariculture, accounts for about a third of all the aquaculture production in the world and has received much attention as a potential supplement for wild-caught marine fisheries. In the present paper, Mariculture is projected as the sunrise industry in India after analysing the current status and existing opportunities.

Keywords: Mariculture, Marine fisheries, Blue economy, Blue revolution

Introduction

The global aquaculture production in 2020 was recorded at 122.6 million tonnes, aquatic animals contributing 87.5 million tonnes and algae 35.1 million tonnes, with a total value of USD 281.5 billion. About one-third of this production comes from farming of the marine species. Asian countries dominated the world aquaculture by producing 91.6 % of the total (FAO, 2022). With 14.16 million metric tonnes production in 2019-20, India is the world's second-largest aquaculture producer next to China, accounting for 7.56 % of global production. It contributes about 1.24 % to the country's GVA and over 7.28 % to the agricultural GVA. During 2019-20, India has exported 12.89 lakh metric tons of fisheries products valued at Rs.46, 662/- crores (USD 6.68 billion). The major farmed species in India are the Pacific white shrimp *Litopenaeus vannamei* and Indian Major Carps (IMC). While issues related to freshwater scarcity and pollution will slow the growth of freshwater aquaculture in coming years, mariculture has been identified as an area of high growth opportunity.

Globally, an estimated two billion people suffer from micronutrient malnutrition and one billion calorie deficient. Thus, providing adequate nutrition to the growing population is a global challenge. Fish currently provides 17% of the world's animal protein, and exceeds 50% in the diets of many least-developed countries (FAO, 2022). One of the most documented nutritional benefits of seafood is the health benefits of unsaturated fatty acids in marine fish and their contribution to brain development, metabolic function, and the prevention of cardiovascular diseases. Seafood also provides essential micronutrients that promote support growth and development, especially in children and pregnant women (FAO, 2022). The proportion of fish eating people in India has been grown from 27.7% in 1987-88 to 50 % in 2015-16. The annual per capita consumption of fish for the entire population was estimated at 5-6 kg, whereas for fish eating population it was found to be 8-9 kg (Shyam, 2016). Assessment of the present trend shows that to meet the increasing demands from the domestic consumers and to cater to the export market, fish production has to come mostly from marine sector only. It is therefore considered that mariculture is the only option for production enhancement from marine sector.

Status of Mariculture in India

Development of mariculture in India began with the farming experiments of seaweeds at Mandapam by the Central Marine Fisheries Research Institute (CMFRI) during early 1970s. CMFRI also took the lead in the development of the technology for edible oyster farming during the 1970s. Intensive research on various aspects of the culture of the Indian backwater oyster (*Crassostrea madrasensis*) has been made and the technology has also been developed for the hatchery production of seed.

CMFRI initiated pearl culture program in 1972 and successfully developed the technology for pearl production in Indian pearl oysters. Success in controlled breeding and spat production of the Japanese pearl oyster *Pinctada fucata* in 1981 and the blacklip pearl oyster *P. margaritifera* in 1984 was yet another breakthrough in mariculture. In *P. fucata*, after implantation of nucleus, lustrous pearls of good quality could be obtained in 6-8 months depending on the size of the nucleus. Later Mabe pearl (image pearl) production was standardised for production of base images with ten different types of moulds. Technology for production of jewellery from Mabe pearl was also standardised by CMFRI.

Two species of marine mussels, the green mussel *Perna viridis* and the Indian brown mussel *P. indica* are the two candidate species for farming in India.

Investigation of the culture possibilities for mussels was also initiated in early 1970s by the CMFRI which resulted in the development of a range of practices for the culture of these species. Kerala was the first state to recognise the advantages of utilizing mussel farming technology in rural development. The Indian mussel production is relatively small and the production is around 10,000 tonnes for the past few years. In India, mussel culture (*Perna viridis*) has become popular in the Malabar area since 1996 following the success achieved by CMFRI in rearing and popularizing green mussel in the backwaters.

In the late 1950s CMFRI had initiated salt water farming of finfishes like milk fish *Chanos chanos* in the coastal earthen ponds at Mandapam. The institute could also achieve success in coastal farming of mullets (*Mugil* spp.), pearl spot *Etroplus suratensis*, tilapia *etc*. Later hatchery facility for finfish breeding and seed production was developed at different places along the Indian coast by CMFRI, Central Institute of Brackishwater Aquaculture (CIBA), Marine Products Export Development Authority (MPEDA), etc. From 2008 onwards breeding and seed production of high valued marine finfishes have been attempted by CMFRI and success has been achieved in seed production of cobia *Rachycentron canadum,* pompano *Trachinotus blochii* and *T. mookalee*, orange spotted grouper *Epinephelus coioides*, pink ear emperor *Lethrinus lentjan* and rabbit fish *Siganus vermiculatus*.

From 2006 onwards, commendable progress in development of indigenous marine cage culture technology has been taken place by CMFRI and National Institute of Ocean Technology (NIOT). Asian seabass, *Lates calcarifer,* cobia, *Rachycentron canadum,* Pompano spp *Trachinotus blochii* and *T. mookalee*, orange spotted grouper *Epinephelus coioides*, *Lethrinus lentjan, giant trevally, Caranx ignobilis, C. sexfasciatus,* red snapper *Lutjanus argentimaculatus*, *L. johnii*, pearl spot, *Etroplus suratensis*, grey mullet, *Mugil cephalus* and milk fish *Chanos chanos* are some of the species successfully farmed in cages in India. Refinement of technologies by CMFRI and other organizations and their adoption has led to rapid spread of cage culture along both the coasts in near shore waters by self-help groups, fisher societies and small-scale entrepreneurs. Technologies currently available in India include seed production and farming of finfishes such as cobia, pompano, sea bass, groupers, snappers, breams and ornamental fishes; shellfishes such as mussels, oysters, clams, Indian white shrimp, green tiger shrimp, blue swimmer crab, pearl oyster, seaweed and ornamental shrimps.

Integrated multi-trophic aquaculture (IMTA) has also been standardised by CMFRI in marine as well as coastal regions in India. Green mussel and seaweeds are integrated successfully with cage farms with additional income, and environmental sustainability.

The current estimated mariculture production in India is about 0.01 million tonne per year, which shows that it is growing in very slow pace compared to global scenario.

Opportunities for Mariculture Development in India

The mariculture sector in India is looked upon as a new horizon, with the potential to play a large role in the Indian economy in the near future. The projected annual mariculture production potential based on area available in the Indian region is 4 to 8 million tonnes. With the vast natural and human resources available, India has the potential and the capacity to establish as a world leader in mariculture. Realizing the sector's potential, the Government has committed to a national target to increase fish production to 22 million metric tons by 2024-2025, which will have a positive impact on 28 million fishers and fish farmers and almost twice that number along fish-related value chains. In India, the inland fisheries production was 10.43 million metric tonnes and marine fisheries production was 3.72 million metric tonnes in 2020. However, the production from mariculture is very meagre now due to various reasons. With many standardized technologies available in mariculture, the opportunities in developing mariculture sector in the country lies in different aspects as given below:

a) **Natural Resource:** The Indian coastline extending up to 8118 kilometres, EEZ of 2.02 million km^2 has a continental shelf area of 0.53 million km^2. With numerous creeks and saline water areas, the opportunities for mariculture are tremendous in India.

b) **Human resource:** As per the 2010 census, there are 3,288 marine fishing villages, 1511 marine fish landing centres in 9 maritime states and two union territories and about 5 million fishermen in India. Mariculture can be taken up with a focus on sustainability through empowering the fishermen by achieving employment generation, social security and increased food and nutritional security and augmenting sea food exports. Many of these wild fish harvesters represent a highly trained workforce that have extensive knowledge of the ocean, boat handling, net mending and maintenance, fish harvesting and quality control that entrepreneurs can easily adapt to their business enterprises.

c) **Educated workforce with excellent animal husbandry skills:** Adequate number of fisheries graduates and specialists in different areas of fish farming (Aquaculture, Nutrition, pathology, ecology, etc.) are available in India to make mariculture sector technically sound.

d) **Domestic and export markets:** It is a major advantage of Indian sub-continent that it has an excellent domestic market for fish and if supply is assured during fishing ban seasons, the returns to the farmer will be very attractive. Export also can be enhanced for fish if adequate post-harvest measures are adopted for post-harvest quality assurance and live fish export.

e) **Strong research and extension capabilities:** In India there are many National Fisheries Research Institutes, equipped with well experienced researchers and infrastructure facilities. Technology development, innovations and transfer can be done at a faster pace with the available expertise in the country.

f) **Strong support by Government of India & Developmental organisations:** The national mission for enhanced fish production by 2024, has been supported by programmes like Pradhan Manthri Matsya Sampada Yojana (PMMSY), which intends to increase aquaculture productivity to 5 tonnes /ha (from the current 3t/ ha), create critical fisheries infrastructure, reduce post-harvest losses, modernise and strengthen value chain, double export earnings to Rs. 1,00,000/crores, enhance domestic fish consumption with corresponding health benefits, attract investments from other sources (State, Private, etc.) and generate direct and indirect employment opportunities of about 55lakhs in fisheries sector from other sources.

The thrust areas identified under PMMSY are:

Mariculture including seaweed cultivation, ornamental and recreational fisheries, catering to demand and supply of quality seed and brood of species and feed (species specific), species diversification through new candidate species, establishment of brood banks, hatcheries, seed rearing units, SPF and SPR seed, genetically improved brood stock and feed mills RAS, biofloc, cage, nano-feed, live feed technology, block chain, value addition, expansion of culture area, etc.

It clearly shows the vast opportunity for mariculture development in India.

Opportunities in Technological Innovations

Technological innovation includes developing new technologies, modifying existing technologies and introducing new technologies to the country.

Opportunities for innovations in different areas are given below:

a) Biotechnological Interventions

The application of genetic principles to increase production from aquatic animals is not yet introduced in India. High valued farmed species can be subjected to genetic improvement programmes for improving reproductive success, growth enhancement, improved flesh quality, disease resistance, etc.

b) Reproduction Innovations

Advancements in genetic breeding represent a promising avenue for ensuring sustainable seafood production in aquaculture. Techniques such as selective breeding, sex reversal, and chromosome manipulation enable the development of strains with enhanced traits such as growth rate, disease resistance, and feed efficiency. Genomic selection and marker-assisted selection facilitate the targeted improvement of specific genetic markers associated with desirable traits. These technologies not only contribute to increased productivity but also aid in the conservation of genetic diversity and the promotion of environmentally responsible aquaculture practices. By harnessing the power of genetic improvement, aquaculture stands to address challenges and meet the growing global demand for seafood while promoting long-term sustainability (FAO, 2018a). Achieving spawning of marine candidate species through environmental manipulation, rather than hormone injection.

Fig. 10.1: Broodstock of orange spotted grouper *Epinephelus coioides* at ICAR-CMFRI

c) Disease Management

Specific and non-specific defence mechanisms of the hosts can be harnessed by incorporating immunostimulants and non-specific immune-enhancers into diets to control diseases in farmed fishes. Use of probiotics as live microbial feed supplements which affect the host animal by improving the intestinal microbial balance to optimise the presence of non-toxic species is also a good option to control diseases.

Development of specific pathogen free (SPF) and specific pathogen resistant (SPR) broodstock for farmed marine species will be useful in biosecurity and mariculture sustainability.

Oral vaccine delivery offers a potential solution to the labor-intensive process of individual hand injections, yet despite progress with mechanical injection systems, the challenge of handling stress persists in animal immunization. Micro-encapsulation, involving the encapsulation of tiny particles or droplets with a coating to form small capsules, and bio-encapsulation, wherein medicines are integrated into living host organisms and subsequently fed to fish, exemplify improved delivery systems with potential applications in various fields.

d) Feed Technologies

Currently, the most heated debates concerning aquaculture development centre on the use of fishmeal and other animal proteins in aquafeeds. Most feeds rely on fish meal and fish oil made from whole fish/ fish parts. But at the rate aquaculture is growing, traditional fish meal supplies aren't likely to keep up with demand.

Plant-based solutions such as: soybean protein concentrates and algae feed are options. Algae have the advantage that it is already a natural part of the food chain for fish. Another option for fish meal replacement is insect-based feed, particularly crickets, mealworms and black soldier fly larvae.

e) Farming Innovations

i) *Cage culture*

Cages are enclosed floating structures made of meshing or nets through which water flows freely. Cages are mostly located in coastal waters and allow fish to benefit from natural ocean currents to deliver oxygen and remove waste. Since net cages are installed in the natural system, fish waste, uneaten feed, and any other waste pose environmental risks to the ecosystems. Net cages are prone to damage due to weather conditions (storms, cyclones or surges) resulting is mass-escape events, resulting in natural ecosystem balance. High density stocking in cages impart challenges such as parasites, disease, growth and breeding deformities, etc. Design innovations (submersible/ submerged cages) in coastal cages and off shore cages (the water quality remains unchanged in offshore conditions) are better innovations for sustainable production.

Fig. 10.2: Indigenous HDPE sea cages

Fig. 10.3: Cobia *Rachycentron canadum* in grow-out cage

Fig. 10.4: Cobia *Rachycentron canadum* harvested from sea cage

Fig. 10.5: Orange spotted grouper harvested from sea cage

Fig. 10.6: Spiny lobster harvested from sea cage

ii) Recirculating Aquaculture Systems (RAS)

Land-based closed-loop farming, also recognized as recirculating aquaculture systems (RAS), integrates fish farming with various aquaculture techniques in a polyculture system, fostering self-sufficient ecosystems by concurrently cultivating animals, plants, and fish. RAS address several environmental challenges by fully separating farmed fish from the larger ecosystem like open sea, eliminating environmental degradation and threats to wild stock. The controlled nature of RAS allows close monitoring and management of health issues under high stocking densities. RAS can be placed anywhere with seawater availability, which is a key sustainability factor for mariculture industry. RAS can offer fresh seafood to regions with limited or no natural access of it, thereby improving health food security and reducing transport-related carbon emissions. RAS offers multiple advantages, including reduced pollution generation, diminished consumption of natural resources, and decreased transmission of diseases and invasive species. For instance, nitrogen waste from fish can be repurposed to cultivate high-quality vegetables, herbs, and plants, contributing to enhanced profitability and sustainability.

Fig. 10.7: Recirculating Aquaculture System (RAS) at Mandapam RC of ICAR-CMFRI

iii) Seaweed Farming

Seaweed cultivation has enormous opportunity in India. About 60 commercially important species are available in the country, which can be farmed at different levels and processed into a myriad of products. It is yet another area of innovation to increase production and investment in seaweeds. The potential of seaweed as a fast-growing and sustainable source of biomass has inspired innovators to develop fertilizers, feed additives, food and feed ingredients, and biofuels from seaweeds. The technology is faced with scaling and seed material availability challenges, but with national and international partnerships among organizations and industries for scaling up initiatives, seed bank establishment, and commercialization of feed and food products rapid industry growth can be expected. Seaweed farming is also one of the potential areas in mariculture, which is expected to open up new avenues for women and weaker sections in the coastal areas to provide income and to promote small, medium or large scale entrepreneurship for the rural unemployed.

Fig.10.8: *Kappaphycus alvarezii* farming by net tube method

iv) Marine Microalgae

Microalgae has a significant role in the nutrition of marine animals in the early life history and consequently in mariculture. Microalgae are utilised as live feed for larval/early juvenile stages of crustaceans (e.g. shrimps, crabs, etc.) and some fish species, for all growth stages of bivalve molluscs (e.g. oysters, clams and mussels) and for almost all zooplankton used in mariculture. Microalgae form sustainable energy sources, through their high biomass productivity and have the ability to remove airborne and water-borne pollutants. They are also potential source of biofuels and bioactive compounds. Commercial mass scale farming of microalgae for any of the above uses will be environmentally sustainable and economically profitable.

v) Marine Ornamentals

The five billion dollar business segment of marine ornamentals includes culture of fish, crustaceans and corals. Commercial marine ornamental culture is relatively a new sector and is often not considered by the mainstream aquaculturists. It could provide considerable profit margins and huge opportunities for coastal communities in India.

Fig. 10.9: Clown fish in marine aquarium

vi) Integrated Multi-trophic Aquaculture (IMTA)

Co-culture of species from different trophic or nutritional levels in the same system with a balance in the biological and chemical processes refers to IMTA. This balance is achieved through the appropriate selection and proportions of different species providing different ecosystem functions. Thus, IMTA results in enhanced production depending on mutual benefits to the co-cultured species and improved ecosystem health, even if the individual production of some of the species is lower compared to what could be reached in monoculture practices over a short term period. The aim of IMTA is to increase long-term sustainability and profitability of mariculture, by converting the wastes of one crop (fish/crustacean) into fertilizer, food and energy for the other crops (seaweed/molluscs). IMTA can be done land-based or in open-water systems, depending on ease of operation and necessity. IMTA has great potential in Indian Mariculture. Integrating seaweed with finfish/ crustacean (lobster/ marine shrimp) the production per unit volume of water can be increased with more income and economic returns. Since IMTA is an ecosystem based culture practice, it will have better acceptance in the light of environmental sustainability. IMTA is a method by which production can be intensified, diversified and yet remain environmentally sustainable. Multi-trophic integration will be a logical innovation in mariculture development in India.

Fig. 10.10: Green mussel *Perna viridis* harvested from cage associated IMTA in Kerala

vii) Artificial Intelligence (AI) and Smart Technologies

Traditional aquaculture is labour intensive and poses many issues for the environment. Therefore, application of smart technologies in aquaculture is aiming to develop the industry in a sustainable manner to enhance production and environment friendly. Using smart technologies, culture operations can be controlled from distance. Farm automation is done by applying Internet of Things (IoT), big data, artificial intelligence, 5G, cloud computing, and robotics. Smart aquaculture can be controlled by robot which can manage facilities, equipment and machineries to operate culture systems to achieve successful production. It can be applied in various system such as cages, hatchery, and breeding, for water quality monitoring, monitoring of cages, and hatchery; optimizing feeding without wastage and adjust feeding time as per requirement, reducing labour by automation etc.

While smart sensors gather data on various parameters, artificial intelligence (AI) can be used in predictive analysis and offer improved data-based decision-making. AI can help in detecting water pollution and alert farmers, when farmers can send out robots to rectify the situation. Current technology demands minimum human intervention, but in the future, most of the decisions will be carried out autonomously.

Apart from AI, augmented reality (AR), virtual reality (VR) and block chain have also made their entry into aquaculture abroad. While the initial investment for these technologies seems to be higher, technology can cut down maintenance costs and improve profit levels from mariculture.

Smart technologies and AI can very well be introduced in mariculture in India for the efficient development of the sector.

Growth Prospects

The growth of Indian seafood business is approximately 15% per year and the demand for seafood is steadily rising around the world. Consumption habits are also changing as a result of increased income and lifestyle changes. As a result, there is potential for small, medium and large scale entrepreneurs who wish to own a firm of seafood business.

Drivers of Mariculture Growth

a) **Demand-side drivers:** Fish is considered as a low-cost, high nutrient, health food source. In India, people are changing their lifestyle and eating habits to include a healthy diet such as marine fish in their diet. This along with rapidly increasing population in the country, the demand for protein also increases proportionally.

b) **Drivers on the supply side:** India accounts for 10% of global biodiversity in terms of fish and shellfish species and the 2.02 million km^2 of EEZ of and the 0.525 million km^2 of continental shelf area provide enormous possibilities for mariculture production. Through farming diversified fish and shell fish species could be produced from different water resources. A lot of new species can be brought under mariculture in the near future.

c) **Policy Support:** Fisheries and Aquaculture Infrastructure Development Fund (FIDF) policy support for INR 7522.48 crores, PMMSY for INR 20,500 crores, Production-Linked Incentive Scheme for the Food Processing Industry for INR 10900 crores and a liberal Foreign Direct Investment (FDI) Policy, where in aquaculture, 100 % FDI is allowed via the automatic route are all excellent opportunities for investing in Mariculture in India.

Further, in 2018-19, Government of India has extended the facility of Kisan credit Card (KCC) to aqua farmers to meet their working capital needs, which includes purchase of seed, feed, and materials required for infrastructure development.

The draft National Fisheries Policy (NFP), 2020, covers the details of National Mariculture Policy (NMP) also with an intent to ensure land management and sustainable development of the resources and associated habitats, along with maintaining the ecosystem integrity meeting the food and nutritional security of the population protecting the sights of fishing and farming communities, building farmers resilience, making Indian fish and fish products globally competitive. Goa is the first state in India to adopt a Mariculture Policy in the country.

Mariculture Interventions in Development

On 25 September 2015, countries adopted a set of 17 Sustainable Development Goals (SDGs) with specific targets to be achieved over the next 15 years, which cover poverty, hunger, gender equality, economic growth and 'life below water'. Though the SDGs are not legally binding, governments are expected to take ownership and establish national frameworks for the achievement of the SDGs through national and sectoral development planning. The aquaculture value chain is able to contribute in achieving SDGs at both national and regional levels. Be it a small, medium or industrial scale operation, mariculture has the potential to contribute to most of the relevant SDGs. Mariculture does provide a regular supply of high quality protein, sparing income for other food and living expenses; it could contribute to good health and education opportunities; generate jobs, and being local, can be undertaken on a part-time basis by women.

Conclusion

A number of benefits will add on to the country with the establishment of a healthy mariculture industry, including wholesome (healthy, nutritious) food to replace that from wild caught fisheries that are declining or at maximum sustainable yield, products for export to improve the nation's balance of trade and foreign currency reserve, enhancement of commercial fishery, economic opportunities for rural and coastal population and new jobs and openings for skilled workers and entrepreneurs.

References

Annual Report (2021-22). Department of Fisheries, Ministry of Fisheries, Animal Husbandry & Dairying, Government of India. 111p.

Barrington, K., Chopin, T. and Robinson, S. (2009). Integrated multi-trophic aquaculture (IMTA) in marine temperate waters. In D. Soto (ed.). Integrated mariculture: a global review. FAO Fisheries and Aquaculture Technical Paper. No. 529. Rome, FAO. pp. 7–46.

FAO (2022). The State of World Fisheries and Aquaculture, 2022. Towards Blue Transformation. Rome, FAO. https://doi.org/10.4060/cc0461en.

Gopalakrishnan, A., Kirubagaran, R., George John, Ponniah, A.G., Gopakumar, G., Sunilkumar Mohamed K., Krishnan , Imelda-Joseph, Abdul Nazar, A.K., Jayakumar, R., Raju, M.S., Sreepada, R.A., Shinoj, P. and Rajesh N. (2019). Draft National Mariculture Policy 2019 (NMP2019); CMFRI Marine Fisheries Policy Series No.17/2020, 22p.

Handbook on Fishery Statistics (2020). Department of Fisheries, Ministry of Fisheries, Animal Husbandry & Dairying, Government of India.176p.

Imelda- Joseph, Gopalakrishnan, A, Vivekanand, B., and Uma, E K (2022). भारत में समुद्री संवर्धन. भारतीय कृषि अनुसंधान परषिद केन्द्रीय समुद्री मात्स्यकिी अनुसंधान संस्थान, Kochi. ISBN 978-93-82263-55-5

Joffre, O. Klerkx, M., L., Dickson, M. and Verdegem M. (2017). How is innovation in aquaculture conceptualized and managed? A systematic literature review and reflection framework to inform analysis and action. Aquaculture, 420: 129-148.

Liu, O. R., Molina, R., Wilson, M. and Halpern B. S. (2018). Global opportunities for mariculture development to promote human nutrition; Peer J., The Open Access Journal for Life and Environ. Res., doi: 10.7717/peerj.4733

Mohan Joseph Modayil (2003). An Overview of the Opportunities in Mariculture in India. Product Development and Seafood Safety (Editors Jose Joseph P.T.Mathew A.C.Joseph V. Muraleedharan), Central Institute of Fisheries Technology (Indian Council of Agricultural Research) Cochin 682 029, India, 413-417.

Rao, G. S., Imelda- Joseph, Philipose, K K and Mojjada, S. K. (2013). Cage Aquaculture in India. Central Marine Fisheries Research Institute, Kochi. 275 p.

Rao, G S. and Imelda-Joseph (2018). Mariculture in India: Status and Potential. In: Aquaculture in India. Narendra Publishing House, Delhi, India, New Delhi, pp. 77-91. ISBN 978-93-87590-36-6

Secretariat of the Convention on Biological Diversity (2004). Solutions for sustainable mariculture - avoiding the adverse effects of mariculture on biological diversity (CBD Technical Series no. 12).

Shyam, S.S. (2016). Fish consumption pattern in India, exports – Overview. Food & Beverage News August 1-15, 25-28p.

Vo, T.T.E, Ko, H., Huh, J.-H. and Kim, Y. (2021). Overview of Smart Aquaculture System: Focusing on Applications of Machine Learning and Computer Vision. Electronics, 10, 2882. https://doi.org/10.3390/ electronics10222882.

11

Trout Culture in Kashmir A Great Opportunity for Sport Fisheries

Inain Jaies*[1], Sameena Khan[2], and Keezia Khurshid[1], Tarannum Mushtaq[2], Imtiyaz Qayoom[2] and Shabir A. Dar[3]

[1]*Division of Aquatic Animal Health Management*
[2]*Division of Aquatic Environmental Management*
[3]*Division of Fishery Engineering*
Faculty of Fisheries, SKUAST-K, Rangil-190006, Ganderbal, Jammu and Kashmir, India

Abstract

Kashmir attracts tourists because of its breathtaking endeavors which include high altitude mountains, alpine lakes and tourist destinations. Among various ventures of tourist fascinations is trout fishing. Popularly known as angler's paradise, natural water bodies including gushing streams, rivers and rivulets accommodate trout fishes to attract tourists all over the world. Trout angling is a crafty sport of chasing fishes where anglers put all his expertise and skills to get them. The UT of J&K, particularly Kashmir, has an ideal climate and geography for raising, breeding, and producing trout fish. Therefore sport fishing has a lot of potential in Kashmir as well and is one of the industries for adventure tourism that is expanding at the quickest. It is important to carefully consider the strategic decision to increase infrastructure for sport fishing tourists.Thus this chapter provides information about the trout culture in Kashmir and how the culture acts as a base for enhancing the sport fisheries.

Keywords: Trout, Sport Fish, Kashmir, Angling, Culture.

Introduction

Trout farming is considered as one of the oldest forms of commercial fish production and dates back to 400 years in Europe and about 150 years in US (Hinshaw, 1990). In India, the upper Himalayan region and Western Ghats are the areas where trout culture and farming are restricted due to suitable climatic conditions. The UT of J&K, especially Kashmir region plays a major role towards trout culture, thus being major contributor of trout as well. In Kashmir, the introduction of trout started in the year 1899 with the arrival of first batch of Brown trout (*Salmo trutta fario*) ova of 10,000 eyed eggs from England (UK). This introduction of trout eggs to Kashmir was assisted by the Duke of Bedford. Woefully, this batch of 10,000 eggs expired on the way due to the unavailability of air transport. Then, the second batch of trout ova arrived in the year 1900 in a magnificent condition from Howeiton in Scotland, assisted by Mr. J.S. Macdonall consisting of 1800 fry. Out of these 1000 swim-up fry were successfully reared up to the adulthood and the first spawning or eggtaking of *Salmo trutta fario*, was done in December 1905 at Harwan near Srinagar (Kashmir) by F. J. Mitchell and rest of 800 fry were reared at his personal residence in Bhag - i - Dilawar Khan in the heart of city near Khankah-i-Moula. In 1905-1906, Mitchell succeeded in establishing a regular Trout Hatchery at Harwan Srinagar Kashmir, having a capacity of 100,000 green eggs. Since then brown trout has established itself throughout the Kashmir and parts of Jammu notably in Baderwah and Poonch providing fly and spoon fishing year after year to thousands of anglers visiting the state.It was from Kashmir that brown trout spread to other sections of the Himalayas. In 1912, Mitchell succeeded in hatching and rearing rainbow trout eggs from a consignment presented to him by the Bristol Water Works England. Today J&K Fisheries Department has achieved remarkable success in trout culture (Hassan and Pandey 2012). In the year 1984, Trout Fish Farming Project was established at Kokernag under cold water fisheries by the assistance of European Economic Community, serving as mother unit for the production of quality seed of Rainbow trout and Brown trout.This project provided a major thrust to the trout fisheries in the UT of J&K from last two decades.Besides this, there is another trout hatchery at Laribal Srinagar where quality trout seed of rainbow trout and brown trout is being produced (Hassan and Pandey 2012).

The common trout species present in the UT are Rainbow trout (*Oncorhynchus mykiss*) and Brown trout (*Salmo trutta fario*). Among these rainbow trout is the most commonly raised species for commercial purpose and brown trout is mainly used for recreational purpose and acts as major fish species for sport fishery. Naturally, trout spawns in cold and well oxygenated streams where bottom contains gravel, this is the reason they are well adapted to the streams of Kashmir. The wild trout takes about three years to mature sexually (Hinshaw *et al.*, 1989). Trout is being cultured in raceways or ponds (with inlet and outlet) and need flowing water. However, some may be produced in cages and recirculatory systems.

Status and scope of trout culture

The climate, topography and environment of UT of J&K, especially Kashmir is well adopted and suitable for rearing, breeding and production of trout fishes. Also the physico-chemical parameters of water are beneficial and productive for trout fishes. The streams of the Kashmir fed with snow and glaciers, mountains, lakes and springs provide boost to the healthy environment for trout fishes. Trout is cold water fish and flourishes well in temperature range of 5°C to 15°C (Hassan and Pandey 2012). The survival rate of trout eggs in Kashmir is about 10%higher than the western average of 46% (Hassan and Pandey 2012). When the trout was introduced in Kashmir by British, at the same time they introduced the system of angler's reserving beats on their favourite rivers. Tourists from all over the world are much attracted to fish angling in Kashmir.

Production Trends in Trout

The trout and salmon are the world's second traded fish species in terms of value accounting 14 percent of global trade following shrimp which accounts 15 percent of the total. Salmon, trout and other diadromous fishes accounts for 6.4 percent (4 million tons) of the total global aquaculture production.More than 206 species belong to the family of Salmonidae, with salmon (*Salmo salar*) and rainbow trout being two of the world´s most important fish species for consumption (Huber *et al.*, 2004).Global salmon and trout production reached 1,112,988 tonnes in the year 2011(Mir *et al.*, 2020). As per FAO estimates the world trout production in 2013 was 8.14 lakh mt. Trout has been introduced to waters on all continents except Antarctica, for entertaining angling and aquaculture practices. In the upland catchments of many tropical and sub-tropical countries of Asia, East Africa and South America, this culture is practiced.The major production centres of trout fish in the world are Europe, Chile, Japan, NorthAmerica and Australia. Chile being world`s leading rainbow trout producer followed by Norway, Iran, Italy and France. In the European Union, the main producer is Italy with 38,000 tonnes followed by France and Denmark with around 33,000 tonnes (FAO, 2013). According to FAO about 8.14 lakh tons of rainbow trout were produced in aquaculture in 2014.There are more than 75 countries that are officially represented in the rainbow trout production statistics, according to FAO FISHSTAT in 2013. Rainbow trout production amounted to above 814 thousand tons, a radical increase from 4400 tonnes in 1950 as per the annual global aquaculture production (Singh *et al.*, 2016). Iran, Chile, Turkey, Norway and Peru, together contribute nearly 64 percent of the total production formed the top trout producing countries back then. vivid rising thump in rainbow trout production in Iran, Turkey, Peru, china and Russia (200 to 1000 percent increase) was observed in the last decade (2003-2013) indicating an important regional shift due to well strategized governance and utilization of available natural resources. Among India's neighbours, China, Nepal and Afghanistan (with annual trout production of 28991, 180 and 150 tons, respectively, in 2013) features in the FAO list of rainbow trout producing countries.

Himachal Pradesh being one of the leading state in rainbow trout farming and seed production in India produced 220.56 tonnes of table sized rainbow trout during 2012-2013. In North Eastern Himalayan region, the production of rainbow trout in Sikkim reached to 80 tonnes during 2013. In peninsular India, Tamil Nadu the total ova production was 2 lakh during 2011 (Rainbow trout farming in India: R& D perspectives bulletin 23 -2015). Annual trout production in India has increased fourfold from nearly 147 to 602 tons, over the past decade (2004-2014) (Singh *et al*., 2017) with principal contributions from Jammu &Kashmir, Himachal Pradesh and Sikkim. However, during the last decade, the annual increase in trout production was 31%. The Production Scenario of Rainbow Trout Aquaculture production is being carried out in several parts of the world (South America, Europe, Asia, Africa and Oceania) outside its native range in North America on International and National level.

The Kashmir province being the leading producer of fish in the state contributes about 80% to the total fish production (Qayoom *et al*., 2015) indicating that about 87% of fish comes from the Kashmir Division (DOF J&K, 2011). The production of trout has increased from 90 tonnes during 2002-03 to 155 tonnes during 2010-11 was more than 450 tonnes in 2016-17 (Annual Report of J&K Fisheries Department) and presenty it is around 1950 tonnes (DOF J&K, 2023). With the establishment of Mother Trout Fish Farming Project at Kokernag in Jammu and Kashmir, trout fisheries received a major thrust during the last two decades under European Union Assistance in 1984. There is another Trout Hatchery of the department located at Laribal where quality Trout seed of Brown Trout and Rainbow Trout is being produced. The projects have the capacity to produce more than 132 lakh eggs per year. Network of more than 42 trout rearing centre in public sector and 513 rearing units in private sector across the State that has been established by the state Department and the number is increasing with time (DoF J&K, 2019).

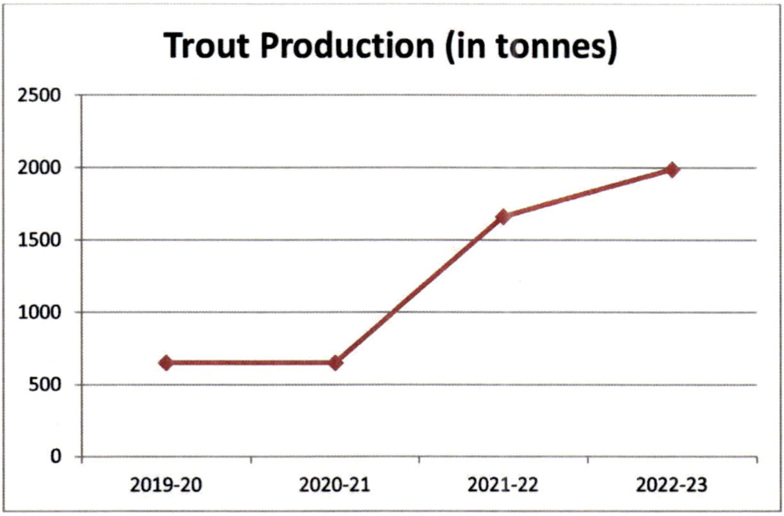

Fig. 11.1: Production trend of trout in Kashmir during last four years

Trout as Sport Fishery

Fishing has a history as long as the planet itself and the oldest known mentions of it stretch back to antiquity. Great Greek thinkers like Aristotle and Plato discussed the activity of angling, while the historian Plutarch provided advice on fishing lines. In the history of sport fisheries, the early development of fishing as enjoyment is not well understood. It is believed that the Norman conquest of England in 1066 brought the first fly fishing enthusiasts to the country (McCully, 2000). One of the earliest books to discuss fishing's worth as a sport is 'Compleat Angler' written by Izaak Walton. It is unclear historically when recreational fishing actually got started.In the United States, fishing is one of the most popular outdoor pastimes. The number of Americans participating in freshwater, sea, and fly-fishing activities increased to almost 55 million in 2020, the greatest number in more than ten years (Statista Research Department, 2022). In Australia, recreational fishing is a popular pastime that is thought to be enjoyed by more than five million Australians. In 1996–1997, 120,000 persons were found to be fishing club members. About 90,000 Australian employments are also supported by recreational fishing. The fishing tackle and bait industry in Australia (with a yearly revenue of over $170 million) and the recreational boating industry (with a yearly revenue of over $500 million, 60% of which is tied to fishing in some form) are the two primary industries involved (Australian Bureau of Statistics, 2003). Less than 1% of all marine catches worldwide are made for recreational purposes.

Regionally, there are differences in trends, with increases in Asia, South America, and Africa, modest declines in Europe and Oceania, and a sharp decline in North America. According to the derived taxonomic composition, Sparidae accounted for 12% of all recent catches, followed by Scombridae (10%), Carangidae (6%), Gadidae (5%), and Sciaenidae (4%) (Freire *et al.*, 2020).

Many medical authorities in the West have long acknowledged recreational fishing, especially angling, as a healthful activity. It has developed as a kind of outdoor activity and gained popularity as "go fishing" while on vacation spurring tourist initiatives. Since trout and mahseer fishing appeals to a variety of tastes and interests, it has become a popular pastime for visitors drawn to the hills of India's northern and southern uplands (Shegal, 1987). Recreational angling has been linked to both social and economic advantages worldwide (Arlinghaus and Cooke, 2009). The national and regional economy have benefited financially from recreational fishing (Everard and Kataria, 2011).

With its extensive coastline, numerous rivers, lakes, and reservoirs, India provides a wealth of chances for sport fishing and angling. In India, the book "Manasollosa" by King Somesvara (1127 AD) on recreational fishing has a reference to angling (Matsyavinoda) (Hora, 1951). Mahseer were initially identified in the Ganges in the early 19th century, and the Oriental Sporting Magazine helped them become well-known as a fishing species (Nautiyal, 2014). Books worth mentioning in this context are 'The Angler in India' on the Mighty Mahseer and 'Circumventing the Mahseer and Other Sport Fishes' (Sehgal, 1987). In streams and rivers of the Himalayas and Western Ghats, British anglers introduced brown trout and rainbow trout for the first time in the 1860s and 1909, respectively (Sehgal, 1999). In order to build upon the earlier roots of sport fishing, established and new angling organisations across the nation committed time and resources (Gupta *et al.*, 2015). With its state and regional officials, the All India Game Fishing Association played a crucial role in promoting sport fishing in India.

The ideal times for sport fishing in India are from October to November and mid-February to mid-May, when the rivers and streams are abundant with all kinds of fish. Fishing is feasible throughout the entire year in India, however, the optimal fishing season may vary due to the varying weather conditions in different locations of India. The optimum months for sport fishing in the Himalayan Rivers are October to November and April to June, while in South India it is April to September. Kashmir, Himachal Pradesh, Uttar Pradesh, and South India are the main locations for recreational fishing in India. The

Indus and Lidder rivers, as well as their tributaries and an extensive network of smaller rivers and streams, are the main locations for fishing in Kashmir. Brown trout are well recognised to inhabit the streams that feed the Beas River in the KulluM anali region of HP. The famous angling sites of India as described by Mandel *et al.*, 2018 are given in table 1.

Table 11.1: Famous angling sites in India

State	Site	Sport Fish
Uttarakhand	Ramganga River	Golden Mahseer
	Saryu rivers	Golden Mahseer
	River Yamuna	Indian trout
Himachal Pradesh	Sangla Valley	Trout
	Manali	Brown trout
Kashmir	Gulmarg	Brown trout
Karnataka	Kaveri River	Mahseer

These days, the rivers Sind and Lider of Kashmir offer some of the best mountain river trout fishing anywhere in the world because the melting of the high altitude glaciers creates the ideal environment for trout throughout the summer. The Sonamarg to Gangabal hike, which passes through some of the most breathtaking Alpine lakes, including Vishansar, Krishansar, Gadsar, and Satsar, is one of the many high altitude lakes that may be reached by fly fishing in Kashmir. When a lake's surface is iced over in the winter, brown trout remain at its bottom and rise to the surface as the ice melts.. According to the department of Fisheries, J&K, every year, the fishing season for trout angling runs from April to September in Kashmir. However, depending on the appropriate conditions that may occasionally exist, the Directorate of Fisheries, J & K Government, may decide to shorten or lengthen the period of angling. The different trout angling beats in Kashmir are:

- Srinagar: Wayil, Wossan, Sindh Middle, Kangan, Cherwan, Mammer, Sumbal, Gund, Rezan, Wangath Stream.
- Anantanag: Aru, Nafran, Lidder, Phrysian, Masjid,Langanbal, Batkote, Bumzoo, Sakras, Wahdan, Pohru, Tricker, Nambal, Kotsu, Langnai,Sarbal, Yenner, Panzeth, Bringhi, Bidhar, Daksum, Dessu, Mather,Nowbugh, Ahlan, Kokernag, Verinag,Ahrabal.
- Budgam: Dood Ganga, Shaliganga.
- Pulwama: Naristan, Lam, Hirpora.
- Kupwara: Mawar, Dringyari, Kalarrooch.

Trout units in Kashmir

There are about four major seed producing hatchery cum breeding units for trout which are located at Kokernag (Anatnag), Laribal (Srinagar), Mammar (Srinagar), Phalni (Rajori) in the UT of J&K (Figure 2).Trout starts to breed in the month of November. The breeding operations are carried out in controlled conditions. For such purpose, the department of fisheries has healthy brooders of parent stock. The eggs of trout fish are obtained by the method of stripping artificially or manually, after segregation of male and female brooders. For the eggs, the fingerlings are raised in hatcheries by mixing the eggs of females with the milt of males and the fertilization of eggs takes place giving rise to fingerlings. Then these fingerlings are supplied to big and small trout rearing units of the UT. These hatcheries, with the assistance from European countries function on scientific and technological grounds. The department has two national fish seed farms, one at Kathua (Jammu) and another at Manasbal (Kashmir). The department of fisheries has exported trout seed to neighbouring states/countries like Himachal Pradesh, Sikkim and Bhutan.

Kokernag

Laribal

Mammar

Phalni

Fig. 11.2: Famous Trout hatcheries in J&K (DOF J&K)

The department of fisheries has at present 32 trout rearing units which are functional. More units are under contruction in newly created districts like Beja

(Baderwah), Aliyalpora (Shopian), Chanser (Kulgam), Krichi (Udhampur), Surasyar, Basantwoder (Budgam), Guzerbal (Bandipora), Dawar (Gurez). In the past one decade not only the new trout rearing units like Uri, Kalantra (Baramulla), Kutlari (Kupwara), Bani (Kathua), Nobra (Leh), Panzeth and Nambal (Anantnag) were constructed but also the existing trout rearing units were upgraded by increasing their rearing capacity and space. In addition the hatching capacity of Kokernag trout farm was enhanced. Also, the various units like Achabal (Anantnag), Khag (Budgam), Jogigund (Anantnag) were rehabilitated and made functional rearing units.

Due to the favourable climatic conditions of J&K UT the trout fish production is very high and mortality rate is low. The trout fish production from both the government and private sectors in J&K has reached to 1663 metric tonnes of which 90-95% is from Kashmir owing to its abundant cold water resources which was about 55 tonnes during 2010-11 and 90 tonnes during 2002-03 (Annual Report of J&K Fisheries Department 2010-11 and 2002-03). It is fact that fish production gradually increased year after year in the UT. Trout fish is highly valued fish in the state due to its nutritive value, less spines and easy to cook. Trout fish has also very high demand outside the UT.

In addition to government sector, private sector of J&K has identified various major thrust areas of overall development of trout fisheries in the UT. Necessary infrastructure shall be raised by the authorities. In this direction the Department of Fisheries is planning to introduce trout culture in private sector for which an additional feed mill is being installed during this financial year. Ministry of Fisheries, Govt. of India during the current financial year released an amount of Rs 18.28 Cr and revalidated an amount of Rs 7.88 Cr under PMMSY and the erstwhile Centrally Sponsored Scheme on Blue Revolution respectively for the construction of local houses of fishermen. Central Govt. also launched PMMSY, a beneficiary oriented flagship scheme during the current financial year under Atma Nirbhar Bharat. The ministry of fisheries during the current financial year released an amount of Rs5.22 Cr out of an approved central allocation of Rs 12.75 Cr mainly for establishment of fish units and the department anticipates establishment of 186 trout units, 37 carp units and 7 biofloc units at an estimated cost of Rs 4.78 Cr (Department of Fisheries, J&K). Establishment of Biofloc units is the first of its kind in the UT as they are based on a new technological intervention where in fish farmers can conserve feed inputs and utilize waste water during production by letting beneficial bacterial colonies proliferate in culture water. During last three decades massive infrastructure in the shape of hatcheries, fish farms, fish rearing units, laboratories etc. have been established. Besides, the establishment of Mega Aquarium at Bahu Fort, Asia's largest trout fish farming project Kokernag, prestigious trout feed mill at Mansbal as well as Hygienic wholesale cum retail fish market at Jammu.

Trout Angling and its Promotion

Jammu and Kashmir serves as the hub for several extracurricular activities. Located in Northern India, it is the perfect vacation spot due to the presence of mountains, waterfalls and the lush valleys. The surroundings, in all their beauty, endow the tourists with a mystical allure. The tourism business makes for roughly 9% of the global GDP (United Nations World Tourism Organization, 2016). Tourists may vary easily engage in a range of outdoor activities such as water sports, hiking, and fishing or angling after taking in their fair share of nature's beauty. Tourists seeking adventure travel are also enhancing choosing sport fishing tourism as an alternative to traditional leisure activities, hence known as Angler's Paradise.The main sport fishing method is angling but the commercial fisheries also employ angling methods like trolling or long line. Anglers are seldom discouraged by the variety of fish that may be found in the streams and steep rivulets.

Over 10% of all new employment created in 2012, according to the World Travel and Tourism Council, were produced in the tourism sector. The industry is anticipated to support (directly or indirectly) over 20.4 million jobs by 2022, which deciphers to an increase in employment of an average of 1% per year over the following ten years in those regions where the impact of crisis on tourism is most noticeable (Magray, 2016). In the tourist sector and in local economies all around the world, recreational fishing plays a significant role. Jammu has the potential to grow its sport fishing industry particularly in terms of *Tor putitora*, *T. tor* and *T. mosal*. Brown trout (*Salmo trutta fario*) and rainbow trout (*Oncorhynchus mykiss*) are the two types of trout that are popular in the Kashmir valley of UT (Bhat *et al*., 2010). Angling is the least expensive way to fish for trout and is permitted to catch on rod and line using both artificial and live baits.The arrival of spring and summer (from the month of April to September) always causes the angling industry in Jammu and Kashmir to boom. Tourists must get permission from the Directorate of the Fisheries, Srinagar, before participating in angling. Numerous adventure-seeking visitors (individuals or groups) go to Kashmir each year for fishing and related activities, mostly for trout fishing. In comparison to other regions in the world, trout fishing in Kashmir is quite affordable.There are several rules that need to be enforced for angling (Magray, 2016):

a) Without a license or permit, angling is not permitted. The license is only valid for a single day.

b) While angling, the license or permission must be carried with the angler and presented to the departmental personnel.

c) It is not permitted to share, sublet, or transfer a fish permit.

d) Fishing with artificial flies is only permitted.

e) No unauthorized individual may work as shikaris (Gillies).

f) Per angler, a single rod is permitted.

g) Permission is for 6-fish maximum per bag.

Besides trout angling, the famous sites of angling in Srinagar are:

- Anchar Lake: In this Anchar Lake, visitors watch and participate in fishing. The visitor is taken through the picturesque sights of the landscape by shikara ride and fishing.
- Dal Lake: The Dal Lake is the magnificent landscape of the Kashmir valley. It is bordered on all sides by gardens and orchards, as well as snow-capped mountain ranges. The best activity here is said to be a Shikara ride. Tourists are seen angling in Jammu and Kashmir from the lovely houseboats and shikaras. The Dal Lake fishing method used by the Kashmiri people is also available for public enjoyment.
- Nagin Lake: Although Nagin Lake is frequently referred to be a distinct lake, it is actually a portion of Dal Lake. This lake is situated at the base of the Zabarwan hills and is encircled by the Hari Parbat on the western side and the Shankaracharya Hill on the southern side. For angling, visitors frequently visit this location.

Promotion of Trout Sport Industry

In 1909, brown trout were introduced to the River Beas in Kullu, Himachal Pradesh. Angling was propagated, conserved, and promoted by the Trout Angling Association in 1914. Almost half a century of fishing has attracted large numbers of anglers. A world-famous wild brown trout habitat is found in the Jibhi and Tirthan streams. A conservation program for Mahaseer and Trout was started by the Himachal Angling Association at Palampur in 1978. Trout Conservation and Angling Association (TCAA), founded in 2010, is constantly working to protect the biodiversity of the valley streams and to restore the environment. During April, Kullu Valley, Himachal Pradesh, hosted a three days All India Open Catch-and-Release Trout Angling Championship. A total of 90 Brown Trout may be caught within two days during three sessions after which length, weight, and other measurements were taken. The winner of the Championship was determined by the size and number of Trout caught. A 5% growth rate per annum is observed in the recreational fishing segment of the tourism industry. Apart from angling, big-game fishing, which is done with speed boats, is another common sport of fishing involving open-water species like Tuna, Sharks, and Marlin. It is also possible to fish with a bow or a crossbow, but this technique is somewhat less common. Knowledge of fish habits, habitats, behaviour, such as migration and foraging, is sometimes essential for using fishing techniques effectively. It is clear that sport fishing has enormous potential in UT as well and is one of the fastest growing adventure tourist markets. It may promote regional economic growth and create job possibilities (Hoitsy*et al.*, 2021).The strategic choice to build up infrastructure for sport fishing tourists should be well thought out. It takes an in-depth knowledge of both the tourist and fishing industries as well as the capacity to forge lasting bonds between them to ensure mutually beneficial links.

Conclusion

The Kashmir Valley is fortunate to have a variety of lakes, rivers, and streams. Despite the fact that recreational fishing has boosted local and national economy, the growing human population has put the water bodies in danger from a variety of anthropogenic pressures. It should be a top priority to improve the governance and management of recreational fisheries, but doing so will require official commitments and a general acceptance of recreational fishing as a legitimate activity. In order to support adaptive management strategies that could result in a thriving and sustainable recreational fishing sector in Kashmir, there is an important need for science, such as historical research and stock assessments.

References

Annual Report of J&K Fisheries Department(DoF J&K, 2019).

Arlinghaus, R., & Cooke, S. J. (2009). Recreational fisheries: socioeconomic importance, conservation issues and management challenges. Recreational hunting, conservation and rural livelihoods: science and practice, 1, 39-58.

Bhat, F. A., Balkhi, M. H. &Yousuf, A. R. (2010). Fish diversity in Kashmir Himalaya: In biodiversity, development and poverty elevation. Dept. of Botany, KU. Pp. 24-27

Everard, M., & Kataria, G. (2011). Recreational angling markets to advance the conservation of a reach of the Western Ramganga River, India. Aquatic Conservation: Marine and Freshwater Ecosystems, 21(1), 101-108.

FAO, 2013. Statistical year book 2013: world food and agriculture. Food And Agricultural Organization of the United Nations, Rome 289.

Freire, K. M. F., Belhabib, D., Espedido, J. C., Hood, L., Kleisner, K. M., Lam, V. W., ...& Pauly, D. (2020). Estimating global catches of marine recreational fisheries. Frontiers in Marine Science, 12.

Gupta, N., Raghavan, R., Sivakumar, K., Mathur, V., &Pinder, A. C. (2015).Assessing recreational fisheries in an emerging economy: Knowledge, perceptions and attitudes of catch-and-release anglers in India. Fisheries Research, 165, 79-84.

Hassan, N. & Pandey, D.N. (2012).Present Status of Trout Fisheries in Jammu and Kashmir. Journal of Pharmacy, 2(5):35-37.

Hoitsy, G., Woynárovich, A. & Moth-Poulsen, T. (2021). Guide to Trout Farming based Angling Tourism.

Hora, S. L. (1951). Knowledge of the ancient hindus concerning fish and fisheries of India. J. Asiat. Soc. Lett, 17, 145-169.

Huber, I., Spanggaard, B., Appel, K. F., Rossen, L., Nielsen, T. & Gram, L. (2004). Phylogenetic Analysis and In-situ Identification of the Intestinal Microbial Community of Rainbow Trout (OncorhynchusMykiss , walbaum) journal of applied microbiology 96(1): 117-132.

Jhingran, V. G. &Sehgal, K. L. (1978).Cold water Fisheries of India, Inland Fisheries Society of India.

Magray, N. A. (2016). Sport Fishing, An Opportunity for Promotion of Adventure Tourism in Jammu and Kashmir. International Journal in Management and Social Science 4(11), 112-123.

Mandal, A., Das, S. K., Biswas, B., & Kairnar, S. O. (2018). Future scope of sport fisheries development in India: a review. Indian Journal of Animal Health, 57(1), 1-16.

McCully, C. B. (2000). The Language of Fly-fishing.Taylor & Francis.ISBN 978-1- 57958-275-3, 76-78.

Mir, S, A., Gul, S., Mir, I, N., and Qayoom, U. (2020). Trout culture in Kashmir: An economical venture for young entrepreneurs. Technical Article .RESEARCHGATE

Mitchell, F. J. (1918). How trout were introduced in Kashmir. Bombay Nat. Hist. Soc. 26(1):176-79 and 295-99.

Nautiyal, P. (2014). Review of the art and science of Indian mahseer (game fish) from nineteenth to twentieth century: road to extinction or conservation?. Proceedings of the National Academy of Sciences, India Section B: Biological Sciences, 84(2), 215-236.

Pandey, N. N., & Ali, S. (2015). Rainbow trout farming in India: R & D Perspectives. Bulletin, 23, 65-74.

Qayoom, I., Balkhi, M.H., Bhat, B.A. (2015). Change in Fishing Patters in Jammu and Kashmir-A Comparative Study. Current world environment, 10(1).

Sehgal, K. L. 1987. Sport fisheries of India.Indian Council of Agricultural Research. New Delhi. pp 126.

Sehgal, K. L. (1999). Coldwater fish and fisheries in the Indian Himalayas: rivers and streams. Fish and fisheries at higher altitudes: Asia. Food and Agriculture Organization of the United Nations Technical Paper, 385, 41-63.

Singh, A. K., Kamalam, B. S., & Kumar, P. (2016).Charting ways to invigorate rainbow trout production in India. Journal of FisheriesSciences.com, 10(2), 0-0.

Singh, A. K., Pandey, N. N., & Ali, S. (2017). Current status and strategies of rainbow trout Oncorhynchus mykiss farming in India. International Journal of Aquaculture, 7.

Sodhi, A. J., Saroch, J. D. & Verma, J. (2013). Fisheries Resources of Kashmir: A case study of River Jhelum. Journal of Chemical, Biological and Physical Sciences 3(2): 1194-1200.

Statista Research Department, 2022 https://www.statista.com/topics/1163/recreational-fishing/#topicHeader__wrapper.

12

Entrepreneurship Opportunities in Aquaponics in India

B.C. Mohapatra, N.K. Chandan and S.K Panda

ICAR-Central Institute of Freshwater Aquaculture, Bhubaneswar - 751 002 Odisha, India

Abstract

This chapter gives an overview of aquaponics that is becoming quite popular with individuals as well as commercial farming entrepreneurs and also discusses the scope and opportunities for entrepreneurship in this field.

Keywords: Aquaponics, Aquaculture, Hydroponics

Introduction

One of the greatest challenges facing the world is how to meet the nutritional needs of a growing human population that is projected to hit 10 billion by 2050. To meet the additional food demands imposed by the nearly 30% population increase, global food production has to increase by as much as 50% (FAO, 2017). This problem is further multiplied due to the imminent threat of an acute shortage of space to grow food, degradation of arable land, limited resources, climate change etc. in traditional ways (Yang and Kim, 2020). Between 1984 and 2016, India faced depletion of arable land at a rate of 0.12 % per year, which means we are left with land to live on, but shortage of food. Hence, to tackle these problems, there is a need to adopt intensive, environmentally friendly, multidimensional and sustainable food production systems like aquaponics (Dijkgraaf *et al.*, 2019).

Aquaponics is a sustainable method of growing fish, and vegetables as well as flowers (Mohapatra *et al.*, 2020). It is becoming quite popular with individuals as well as commercial farming entrepreneurs, and governments interested in modern farming for year-round food production in different climatic conditions. Moreover, with this type of versatile farming, we can grow considerably more food with less water, land and labour than the traditional farming systems.

In recent years aquaponics is gaining public interest in urban as well as peri-urban areas, in line with the trend towards multidimensional productivity with limited resources. Now this boasts diving it from small-scale operation to the brink of commercialization. Marine, inland saline, and decoupled aquaponics has increasingly gained popularity in last decades. However further expansion is demanded as a way-forward to tackle the global challenges (Verma *et al.*, 2023).

Principle of Aquaponics

Aquaponics is a form of farming that unites raising fish in tanks (recirculating aquaculture) with soilless plant culture (hydroponics). In aquaponics, the nutrient-rich water from the fish supplies natural fertilizers for the plants, and the plants help to purify the water for the fish to survive. Traditionally, in aquaculture practices, the nutrients provided by the fish in the form of faecal matter and feed waste would normally be discharged and could cause environmental pollution. Also, in hydroponics, additional nutrient solutions are required for the growth of plants. A cumulative approach to solve both of these problems is the removal of fish-generated waste and converting it to nutrients for the plants, which prolongs water use; and minimizes water discharge outside of the system and hence reduce the waste of resource and provides an economic benefit from fish and plant production in a sustainable manner (Joyce *et al.*, 2019; Nuwansi *et al.*, 2019).

Some Major Attributes of the Aquaponics System

Aquaponics system has many advantages over conventional crop production systems:

- The wastes of one biological system are used as inputs for another biological system.
- The integration of fish and plants increases the diversity and enhances system stability and sustainability.
- Reduces the amount of waste discharged to the environment and minimizes the negative environmental impacts via water filtration and removal of nutrients from the cultured water before it leaves the system.
- Minimizes water wastage.
- The aquaponics system can be operated in different climatic conditions.

- In arid areas, where water is scarce, aquaponics is an appropriate technology that allows the production with the reuse of water.
- Fewer pests since we are operating the system indoors or in controlled conditions.
- No headache of weeding.
- For the commercial growers, aquaponics provides two streams of income, fish and veggies, rather than just one.
- Aquaponics farm does not have a need of farmland with fertile soil, or even land with soil. Aquaponics can be constructed in any type of land profile or in rooftops.
- System is simple and easy to adopt.

Limitations

- High initial start-up costs.
- Requires knowledge not only on growing vegetables, but also on fish culture.
- Mistakes in managing the system can cause its collapse.
- System needs regular management.
- Continuous energy demand.
- Requires proper knowledge on variety of fish and plant that can be used in the system and their numbers.
- Proper hydrodynamics throughout the re-circulatory system is a strict mandate.

Types of Aquaponics

i. Media Bed Culture (MBC)

This is the simplest form of aquaponics that makes use of containers filled with suitable growing media such as expanded clay balls (hydroton), pumice stone, gravel or something similar materials to support the roots of plants. The bed is flooded and drained of nutrient-rich water to give the plants the nutrients they need. The growing media used to support the plants plays role of both mechanical and biofilter to confine and break down wastes. This is economical, simple to put together and productive on a small scale. Because the media supports plants like soil, we can produce large root mass plants such as fruits, flowering plants and vegetables. As because the media is not space-efficient, it requires considerably more inputs of labour and media bed designs, thus, hard to scale into commercial use.

ii. Deep Water Culture (DWC)

Also recognized as raft or float systems, works on the idea of floating plants on top of the water allowing the roots to hang down into the nutrient-rich water. This can be done in several ways and this method is one of the more commonly practised commercial methods. The method can be utilized in various ways such as floating foam rafts on top of freshwater bodies or pumping water from the fish tank to floating rafts with plants floating on water surfaces. This type of system is well suited to the ponds with eutrophication.

iii. Nutrient Film Technique (NFT)

NFT aquaponics shows potential for custom aquaponics designs. These aquaponic designs are popular in the commercial industry because of their space efficiency and lower labour costs. In NFT systems, water is lifted from the biofilter into each hydroponic pipe or tank with a small amount of flow creating a shallow stream of nutrient-rich water flowing along the bottom. Plants sit in small plastic cups allowing their roots to access the water and absorb the nutrients. This system uses less water than the other two methods.

Components of the Aquaponics System

Aquaponics consists of two main parts, with the aquaculture part for raising aquatic animals and the hydroponics part for growing plants. Although consisting primarily of these two parts, aquaponics systems are having several components responsible for the effective removal of solid wastes, for adding bases to neutralize acids, or for maintaining water oxygenation.

Typical components of an NFT Aquaponics include

Rearing tank: The tanks are for raising and feeding the fish. Tanks can be manufactured with different materials depending upon the scale of operation. The main preferable materials are polyliner, FRP, PP, HDPE and in some cases RCC.

Fig. 12.1: Fish Rearing Tank

Settling chamber: A sedimentation unit for large chunks of uneaten food, detached biofilms and fish excreta of particulate size larger than 100 μm.

The water from the fish culture tank containing aquaculture waste particles having various SG flows into the settler. With decreased turbulence and flow rate, the heavier solids (SG>1) settle at the bottom and the suspended ones (SG≤1) are captured by an open-cell polyurethane foam layer supported by an FRP mesh. The filtered water is then fed into the biofilter, where the nitrification takes place. The polyurethane foam in the settling tank was cleaned every day to remove any particles trapped which can lead to pathogenic microbial growth.

Biofilter: The unit responsible for the nitrification bacteria colonization and biological conversion of ammonia to nitrate and nitrates ions usable by the plants.

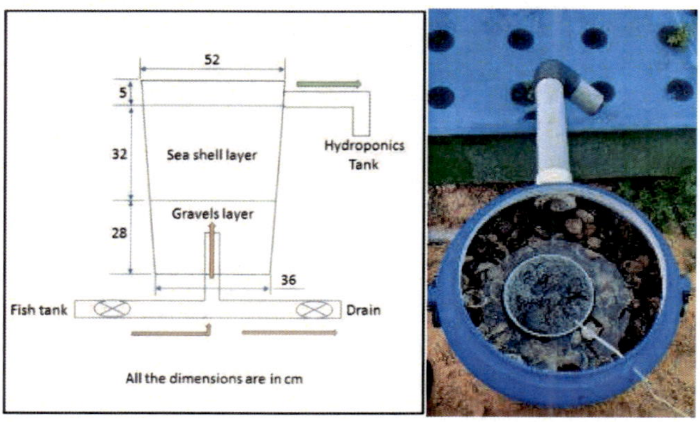

Fig. 12.2: Biofilter Layout

Hydroponics tank: The portion of the system where plants are grown by absorbing excess nutrients from the water.

Fig. 12.3: Hydroponics Tank

Sump: The lowest point in the system where the water overflowing from the hydroponics tank is stored and can be pumped back to the rearing tank automatically/manually.

Sump Automated Sensor System

Fig.12.4: Sump and Automated sensor system

Depending on the sophistication and cost of the aquaponics system, the units for solids removal, bio-filtration, and/or the hydroponics may be combined into one unit, which prevents the water flowing directly from the aquaculture part of the system to the hydroponics part.

Live Components

An aquaponic system depends on different live components to work successfully. Those are plants, fish (or other aquatic animals) and bacteria. Some systems also include additional live components like worms. The ratio of fish to plant is a critical aspect of aquaponics since the balance between the two results in the absorption of nutrients by plants for optimum plant growth and plays a conclusive role in the economic success of the aquaponic system (Palm *et al.*, 2019).

Fish

Freshwater fish are the most common aquatic animals raised using aquaponics (although freshwater prawns are also sometimes used). Among the various fish cultured in aquaponics, tilapia is the model species that is widely cultured in aquaponics, followed by carps and catfishes (Yep and Zheng, 2019). Followings are the common groups of fishes experimented in the aquaponic system.

Aquarium fish, Tilapia, Yellow perch, Trout, Catfish, Bass, Bluegill and Carps.

Plants

The plants that are compatible with hydroponic system will also be compatible with aquaponics systems. The plants like leafy green vegetables, flowers and herbs are most commonly and most successfully grown in the aquaponics system. The herbs are the best choice on account of their rapid growth, high market price, and higher income per unit area per unit time in the aquaponic system. (Rakocy *et al.* 2006). The plant selection primarily depends on the fish stocking density and the nutrient concentration in the culture water (Gosh and Chowdhury, 2019). Following plants are suitable for the aquaponic farming.

Spinach, lettuce, Swiss chad, chicory, basil, tomato, okra, pepper, capsicum, marigold, etc.

Optimization of Stocking of fish and plant

Optimizing of stocking density or determining of ideal fish-to-plant ratio is crucial in aquaponics culture as it influences the system's productivity and profitability (Nuwansi *et al.*, 2021). The fish shows a species-specific response to stocking density in the system (Maucieri *et al.*, 2019). An ideal stocking density ensures a balance between the nutrients from fish culture and the nutrient uptake and assimilation to plant biomass (Nuwansi *et al.*, 2021). Hence, optimum stocking density is a key factor for designing an aquaponic system for optimum production, water quality, growth performance, and yield of fish and plants in the aquaponic system (Al Tawaha *et al.*, 2021; Nuwansi *et al.*, 2021).

Hydraulic loading rate and water flow in aquaponics

Water flow and hydraulic loading rates are significant for optimizing the performance of an aquaponic system (Yang and Kim, 2020). The appropriate water flow rate in aquaponics ensures optimal water quality and thereby improves the efficiency and production of the system (Peter *et al.*, 2021). Discontinuous flow provides uniform distribution of nutrients and enhances aeration in the system during the water circulation whereas, the continuous flow ensures higher water retention and thereby increased contact time between microbes and plant roots; hence yield will be higher with better production and feed efficiency in the system (Maucieri *et al.*, 2018).

Nutrient supplementation

One of the key challenges in conventional aquaponics is the difference in the nutrient requirement in the two components, fish cultured and hydroponics (Goddek *et al.*, 2019). The aquaponic system, which exclusively relies on fish feed as the nutrient source for the plant, is found to be deficient in some nutrients like K, Fe, Ca, etc., which are required for optimal plant production (Angkha *et al.*, 2020). The sub-optimal availability of the nutrients like K, Fe, etc., results in yield reduction in the system. Thus, fertilizer supplementation is required *via.* foliar application (Roosta, 2014) or into culture water/ fertigation (Rakocy *et al.*, 2006) or through supplementation in fish diet (Siqwepu *et al.*, 2020a, 2020b) to prevent deficiency and thereby enhance the plant production in the system.

NFT Aquaponics System Developed by ICAR-CIFA

i. NFT Aquaponics System in Grow Pipes

An aquaponics system developed using PVC pipes as hydroponics nutrient film channels and circular Fiber Reinforced Plastic (FRP) tanks with water holding capacity of 2000 L for fish culture (Anantharaja *et al.*, 2017). The height of the FRP tank is 0.9 m. The water from the fish culture tank is pumped for 5 min in every 2 h over the trickling filter bed using a submersible water pump having the timer. The volumetric flow rate (Q_{MAX}), Head (H_{MAX}) and Wattage of the pump used are 2000 L/h, 2.5 m and 55W, respectively. A trickling filter is developed using gravel ranging from 10 mm to 25 mm in a plastic bucket. The water from the fish tank is pumped into the filter bed. While the water passes through the gravels in the trickling filter, the organic material present in the water is metabolised by the microorganisms attached to the medium. The outlet pipe of trickling filter is directly connected to the plant grow pipe using 1" PVC pipe. NFT grow pipes are made using 6-inch dia. PVC pipes. Four parallel NFT grow pipes each having length of 3 m and 9 perforations (3-inch diameter each) for holding net pots of 3-inch size are employed for growing plants. The NFT grow pipes are connected using 6" 'L' joint. The water coming out of the trickling filter enters into the first NFT grow pipe and passes through all the NFT pipes and returns to the fish culture tank through a 1" PVC pipe connected at outlet provided in the fourth NFT pipe. A Slope of 1 cm is allowed in the NFT pipes for gravitational flow of water.

Fig.12.5: NFT Aquaponics System in Grow pipes

Fig.12.6: NFT Aquaponics Set-up at ICAR-CIFA

ii. NFT Aquaponics System in FRP Tanks

A portable NFT aquaponics system has been designed, developed and tested at ICAR-CIFA, Bhubaneswar (Mohapatra *et al.*, 2020, 2021). It consists of four major components, such as FRP round fish culture tank (ø2.15 × 0.9 m) with operational capacity 2800 L, biofilter unit made up of Polypropylene (PP) of 100 L capacity, FRP rectangular hydroponics tank (4 × 0.9 × 0.35 m) having 2.64 m^2 plantation area and High-density Polyethylene (HDPE) sump (ø0.6 × 0.7 m) of 200 L capacity. Implementation of custom designed and calibrated automatic water recirculation system gives an average flow rate of 94.7 L/h for continuous flow of nutrients from fish culture tank to hydroponics tank. The designed system harnesses gravity flow in 75 % of the cycle. The efficiency of the whole system is mainly focused on the effectiveness of the biofilter in the chemical conversion of Total Ammonia Nitrogen (TAN) which shows 61.97 % reduction with the hydraulic loading of 3.68 LPM of filter media.

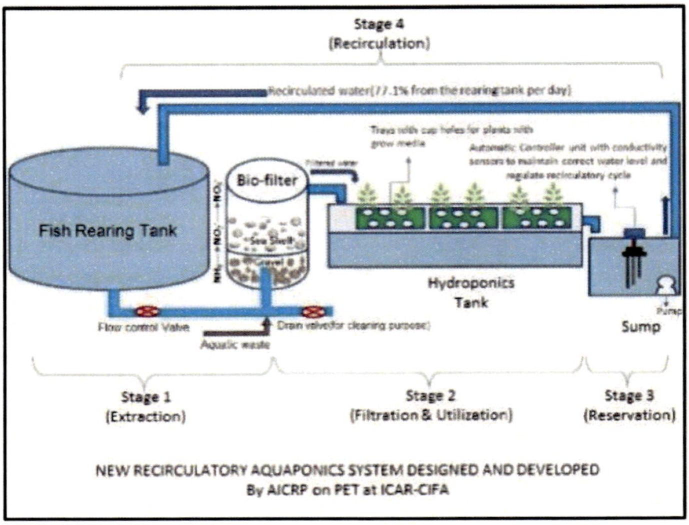

Fig.12.7: NFT Aquaponics System in FRP Tanks

iii. Vertical Aquaponic system

A glass rectangular tank (0.6 × 0.46 × 0.46 m) for 100 L operational water volume for fish rearing was used for the system (Mohapatra *et. al.*, 2023). The side walls are 5 mm thick annealed glass. The vertical plantation pipe is an Ø160 mm PVC pipe with a 4 kg/cm^2 load capacity. The height of the pipe is 1.37 m. This pipe being thermoplastic, the plantation holes are made by heating and reshaping the angle to 20° for sufficient support to the plant stem and water retention. 76.2 mm neck diameter, 50.8 mm base diameter and 63.5 mm hydroponics cups are used for plantation in this experiment as they provide an adequate volume of ~ 204 CC to have the clay balls as grow media with surface area of nearly about 195 cm^2 required for an average ornamental plant for indoor purposes.

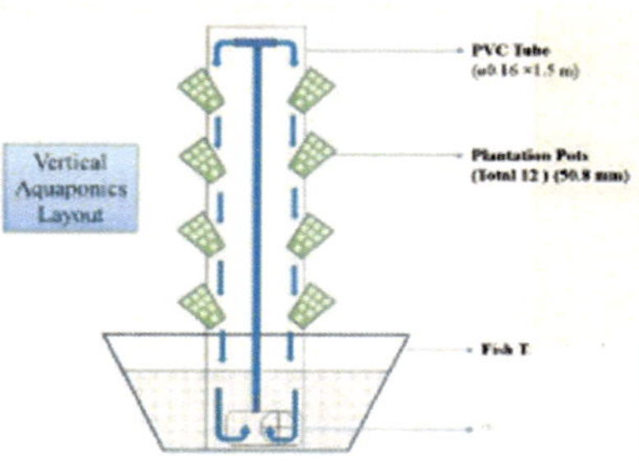

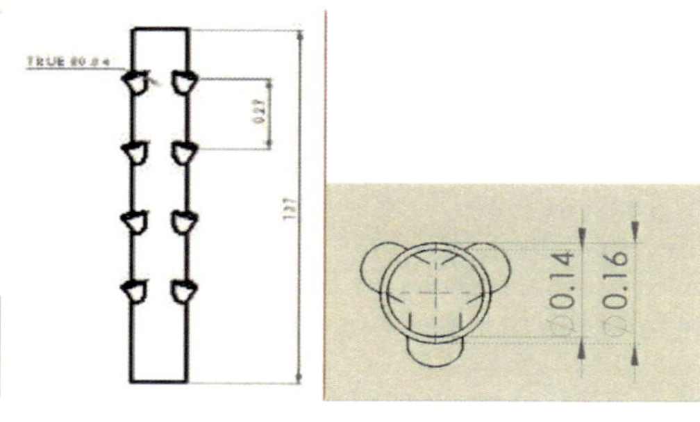

Layout of vertical aquaponics

Engineering designs of plantation pipe

Fig. 12.8 Vertical Aquaponic System

Biological Species Tested in Aquaponics System at ICAR-CIFA

Using these systems, various plants like poi saga (*Basella alba*), marigold (*Tagetes erecta*), bottle gourd (*Lagenaria siceraria*), lady's finger (*Abelmoschus esculentus*), cherry tomato (*Solanum lycopersicum* var. cerasiforme) and long bean (*Vigna unguiculata*) have been grown along with different fish species at ICAR-CIFA (Anantharaja *et al*., 2017). The fish species experimented on are climbing perch (*Anabas testudineus*), tilapia (*Oreochromis niloticus*), pangas (*Pangasius hypophthalmus*) and silver barb (*Puntius gonionotus*). Rearing silver barb seed and marigold plant in it the net profit is calculated to be INR 498 per m^2 of land in a year (Mohapatra *et. al*., 2021).

Current Stage of Aquaponics Experimentation

Various aquaponics farming techniques like NFT and DWC aquaponics in pilot scale experimental setups were designed and developed at ICAR-CIFA, Bhubaneswar campus under ICAR-AICRP on PEASEM. Multiple preliminary trials with different species of fish like puntius, tilapia and IMC and plants like lettuce, long beans, green chili, okra, tomato, marigold, etc. were tested in the developed setups and results were published. The portable NFT aquaponics system developed using FRP (Fiber Reinforced Plastic) has shown good economics and sustainable results in the present scale. The system was designed with scalability in mind. With availability of adequate funds and infrastructure, the present plastic based system can be scaled up to meet the need of the marginal as well as industrial farmers/ entrepreneurs.

Vertical aquaponics is also one emerging technique which can further reduce space utilization by adopting a vertical plantation fashion. Some promising experimental results were also published by ICAR-AICRP on PEASEM center operating at ICAR-CIFA, Bhubaneswar. Some of the remarkable achievements by the firms throughout the country are also seen. Some of them are: UrbanKissan (operating in the cities of Hyderabad and Bangalore), UGF Urban Green Fate (Mumbai-based vertical firm, which was founded in 2012 in the US, and utilises vacant plots, gaps between homes, buildings and restaurants for the purpose), Triton Foodworks (firm incorporates hydroponics and aeroponics at their Delhi based vertical firm), 365Dfarms (Pune based firm is the first to develop movable hydroponic vertical farm), Madhavi Farms (Bangalore, Karnataka), etc.

Aquaponics in Business Creation

Public interest in commercial farming of aquaponics has increased dramatically in recent years, in line with the trend towards more integrated value chains, greater productivity and less harmful environmental impact compared to other production systems. New business models are opening up, with new customers and markets, and with this expansion comes the potential for confusion, misunderstanding and deception. New stakeholders require guidelines and detail concerning the different system designs and their potentials (Palm *et al.*, 2018).

In the country, nowadays a large group of novice farmers is emerging with a keen interest of mechanised or computer aided farming techniques. The environmental consciousness and tech savvy approach is also pointing towards a sustainable yet modernized farm production system. Aquaponics is helping these farmers to safely plunge into the vast verse of agriculture while giving the country advances in farm mechanization, organic farming and utilization of renewable energy for horticulture. Some of the examples of the commercial scale aquaponics farms in India are, Madhavi Farms of Bengaluru, Red Otter Farms of New Delhi and Anjali Aquaponics Farm of Hyderabad, etc. These farms have been operating with sustainable commercial produce from aquaponics with an average ROI of 7-9 years. Therefore, a well-designed setup and proper knowledge of the crops can help entrepreneurs to adopt aquaponics. The Government of India has also been supportive through schemes like PMMSY for adequate subsidies to the entrepreneurs to promote such technology.

Since the interest in aquaponics is increasing, the major challenge is to do it feasible and reliable at commercial scale. The concept of precision farming usually applied in the traditional farming sense is now being introduced, leading to the need to adopt sensing, smart and IoT systems for monitoring and control of its automated processes (Yanes et al., 2020).

Conclusion

- The use of lightweight FRP material made the system much more convenient for transportation to various remote locations and considered eco-friendly.
- The developed system can be a useful tool for hobbyists and also can add recreation to common users.
- Indian socio-economic condition would be greatly benefited by the new design which is aiming for the livelihood improvement of small and medium-scale farmers as well as the people in remote areas by the production of healthy food for the local market.

Future prospects

Aquaponics systems developed under ICAR-AICRP on PEASEM at ICAR-CIFA, Bhubaneswar research facility are in pilot scale for marginal farmers. However, the designs are easily scalable which can fulfil the farming demands as per the need. Aquaponics is a novel technology that holds great potential to overcome the challenges in the aquaculture and agricultural sector. Aquaponics being a multidimensional culture system, this technique has the advantages over traditional farming systems for optimum utilization of resources. The higher consumption of natural resources with growing global population, the aquaponics, aeroponics and hydroponics can meet the future demands.

Acknowledgement

The authors are thankful to the Indian Council of Agricultural Research - All India Coordinated Research Project on Plasticulture Engineering & Technology (AICRP on PET), Ludhiana, Punjab, India for financial support and Director, ICAR-Central Institute of Freshwater Aquaculture, Bhubaneswar, Odisha, India for providing facilities for the experiment.

References

Al Tawaha, A.R., Megat Wahab, P.E., Binti Jaafar, H., Kee Zuan, A.T., Zafri Hassan, M., Al-Tawaha, M.A.R., 2021. Yield and nutrients leaf content of butterhead lettuce (Lactuca sativa) in response to fish nutrient solution in a small scale of aquaponic systems. Ecol. Eng. Environ. Tech. 22 (6), 85–94.

Anantharaja, K., Mohapatra, B. C., Pillai, B. R., Kumar, R., Devaraj, C., & Majhi, D. (2017). Growth and survival of climbing perch, Anabas testudineus in Nutrient Film Technique (NFT) Aquaponics System. Int. J. Fish. Aquat. Stud. 5(4), 24-29.

Angkha, B., Verma, A.K., Kumar, S.H., Prakash, C., Thomas, R.M., 2020. Mobilization of mica by Bacillus sp. and its effect on Nile tilapia (Oreochromis niloticus) cum holy basil (Ocimum tenuiflorum)–based aquaponic system. Aquaculture Int. 28 (5), 2045–2058.

Dijkgraaf, K.H., Goddek, S., Keesman, K.J., 2019. Modeling innovative aquaponics farming in Kenya. Aquaculture Int. 27 (5), 1395–1422.

FAO. 2017. The future of food and agriculture – Trends and challenges. Food and Agriculture Organization, Rome. ISBN 978-92-5-109551-5.

Goddek, S., K¨orner, O., 2019. A fully integrated simulation model of multi-loop aquaponics: A case study for system sizing in different environments. Agric. Syst. 171, 143–154.

Gosh, K., Chowdhury, S., 2019. Review of Aquaponics System: Searching for a Technically Feasible and Economically Profitable Aquaponics System. Journal of Agricultural, Environmental and Consumer Sciences 19, pp. 5–13.

Joyce, A., Goddek, S., Kotzen, B., Wuertz, S., 2019. Aquaponics: closing the cycle on limited water, land and nutrient resources. In: Aquaponics Food Production Systems. Springer International Publishing, Germany, pp. 19–34.

Maucieri, C., Nicoletto, C., Junge, R., Schmautz, Z., Sambo, P., Borin, M., 2018. Hydroponic systems and water management in aquaponics: a review. Ital. J. Agron. 13(1), pISSN 1125-4718.

Maucieri, C., Nicoletto, C., Zanin, G., Birolo, M., Trocino, A., Sambo, P., Borin, M., Xiccato, G., 2019. Effect of stocking density of fish on water quality and growth performance of European carp and leafy vegetables in a low-tech aquaponic system. PLoS One 14 (5), e0217561.

Mohapatra, B. C., Chandan, N. K., Panda, S. K., Majhi, D., & Pillai, B. R. (2020). Design and development of a portable and streamlined nutrient film technique (NFT) aquaponic system. Aquacultural Engineering, 90, 102100.

Mohapatra, B. C., Panda, S. K., Chandan, N. K., Majhi, D., 2023. Design and Development of User-Friendly Vertical Aquaponics Set-up for Ornamental Fish and Plants. Curr. World Environ. 18(2), 515-523.

Mohapatra, B.C., Prabhati K. Sahoo, N.K. Chandan, M. Mahapatra, S.K. Panda, D. Majhi & B.R. Pillai, 2021. Rearing of fish, Puntius gonionotus seed and marigold, Tagetes patula plant in NFT aquaponics system for economic returns. Biological Forum - An International Journal, 13(3): 444-449.

Nuwansi, K.K.T., Verma, A.K., Chandrakant, M.H., Prabhath, G.P.W.A., Peter, R.M., 2021. Optimization of stocking density of koi carp (Cyprinus carpio var. koi) with gotukola (Centella asiatica) in an aquaponic system using phytoremediated aquaculture wastewater. Aquaculture 532, 735993.

Nuwansi, K.K.T., Verma, A.K., Rathore, G., Prakash, C., Chandrakant, M.H., Prabhath, G. P.W.A., 2019. Utilization of phytoremediated aquaculture wastewater for production of koi carp (Cyprinus carpio var. koi) and gotukola (Centella asiatica) in an aquaponics. Aquaculture 507, 361–369.

Palm, H.W., Knaus, U., Appelbaum, S., Strauch, S.M., Kotzen, B., 2019. Coupled aquaponics systems. In: Goddek, S., Joyce, A., Kotzen, B., Burnell, G.M. (Eds.), Aquaponics Food Production Systems. Springer International Publishing, Germany, pp. 163–199.

Palm, H. W., Knaus, U., Appelbaum, S., Goddek, S., Strauch, S. M., Vermeulen, T., Jijakli, M.H. & Kotzen, B., 2018. Towards commercial aquaponics: a review of systems, designs, scales and nomenclature. Aquaculture Int., 26(3), 813-842.

Peter, R.M., Verma, A.K., Saharan, N., Tiwari, V.K., Chandrakant, M.H., Thomas, R.M., 2021. Effect of hydraulic loading rate on production of tomato (Solanum lycopersicum) with pearlspot (Etroplus suratensis) in recirculating aquaponic system. J. Agric. Eng. 58 (3), 300–308.

Rakocy, J.E., Masser, M.P., Losordo, T.M., 2006. Recirculating Aquaculture Tank Production Systems: Aquaponics- Integrating Fish and Plant Culture, Vol 454. SRAC Publication - Southern Regional Aquaculture Center, pp. 1–16.

Roosta, H.R., 2014. Effects of foliar spray of K on mint, radish, parsley and coriander plants in aquaponic system. J. Plant Nutr. 37 (14), 2236–2254.

Siqwepu, O., Salie, K., Goosen, N., 2020a. Evaluation of chelated iron and iron sulfate in the diet of African catfish, Clarias gariepinus to enhance iron excretion for application in integrated aquaponics systems. J. World Aquacult. Soc. 51 (4), 1034–1053.

Siqwepu, O., Salie, K., Goosen, N., 2020b. Evaluation of potassium diformate and potassium chloride in the diet of the African catfish, Clarias gariepinus in a recirculating aquaculture system. Aquaculture 526, 735414.

Verma, A. K., Chandrakant, M. H., John, V. C., Peter, R. M., & John, I. E., 2023. Aquaponics as an integrated agri-aquaculture system (IAAS): Emerging trends and future prospects. Technological Forecasting and Social Change 194, 122709.

Yanes, A. R., Martinez, P., & Ahmad, R., 2020. Towards automated aquaponics: A review on monitoring, IoT, and smart systems. Journal of Cleaner Production, 263, 121571.

Yang, T., Kim, H.J., 2020. Effects of hydraulic loading rate on spatial and temporal water quality characteristics and crop growth and yield in aquaponic systems. Horticulturae 6 (1), 9.

Yep, B., Zheng, Y., 2019. Aquaponic trends and challenges – a review. J. Clean. Prod. 228, 1586–1599.

Yep, B., Zheng, Y., 2020. Potassium and micronutrient fertilizer addition in a mock aquaponic system for drug-type Cannabis sativa L. cultivation. Can. J. Plant Sci. 101 (3), 341–352.

13

Entrepreneurship Opportunities in Fish Craft and Gear Technology

B. Meenakumari

Former Deputy Director General, ICAR (Fisheries) and Chairperson, National Biodiversity Authority Mangalendu, Priyadarshini Nagar Thevara, Cochin - 682013, Kerala, India

Abstract

This chapter deals with fishing crafts and gears and throws light on the scope and entrepreneurship opportunities in fish craft and gear technology for sustainable fisheries and aquaculture.

Keywords: *Fish craft and gear technology, Fishing gears, Boats, Deep-sea fishing*

Fisheries: A glance

The fisheries sector is a complex enterprise that includes capture and culture-based fisheries, including mariculture, coastal aquaculture, freshwater aquaculture, and cold-water fish culture that contribute to the country's food, health, economy, exports, employment, and tourism. Fisheries and aquaculture continue to provide millions of people with food, nutrition, revenue, and a means of subsistence. Fish is an inexpensive and high-quality source of animal protein, making it an excellent choice for reducing hunger and malnutrition in the country. In 2020, global aquaculture production reached a new high of 122.6 million tonnes, including 87.5 million tonnes of aquatic animals valued at USD 264.8 billion. Inland aquaculture produced 54.4 million tonnes, while marine and coastal aquaculture made 68.1 million tonnes. Over time, the Indian fishing industry evolved and became a significant socioeconomic asset for the country (FAO, SOFIA, 2022)

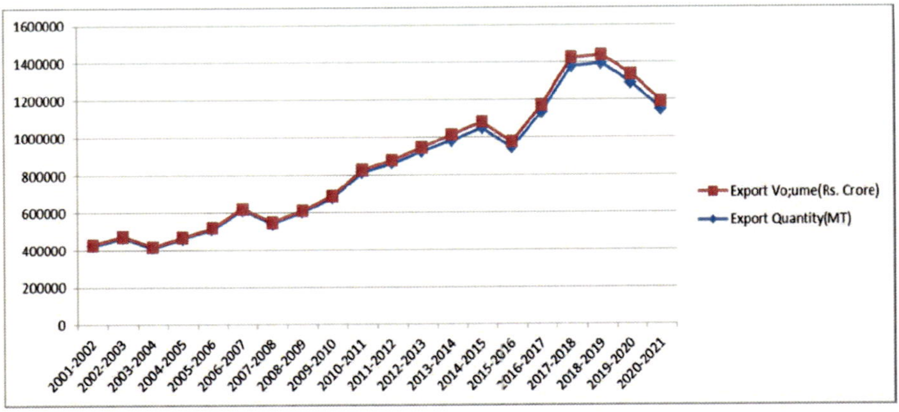

Fig. 13.1: India's Export growth of Marine products
Source: Sipra Karmakar

India is the world's third-largest producer of fish and aquaculture, accounting around 16% of total inland and 5% of total marine fish production. In 2021-22. The overall marine and inland fish production in India was 162. 48 lakh tonnes, with 121.21 lakh tonnes and 41.27 lakh tonnes coming from the inland and marine sectors, respectively. The fishing industry is vital to the national economy and is a major contributor to the country's foreign exchange gains. India's fish production has risen from 56.56 lakh tonnes in 2000-01 to 162.48 lakh tonnes in 2021-22.The majority of fish produced is for human consumption, and this proportion is predicted to increase from 89% in the base year (2016-18) to 91% by 2028 (OECD/FAO, 2019). In 2021-22, frozen fish, cuttlefish, and squid contributed 6%, 3%, and 5% of total marine product export value, respectively.

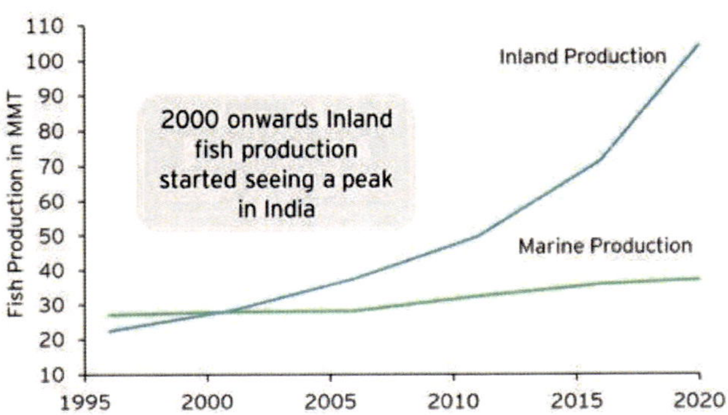

Fig. 13.2: Fish production in India
Source: Department of fisheries, GOI, National portal

Our current annual fish production on capture and culture fisheries is 16.25 million tonnes, and it provides a low-cost source of protein that contributes to our people's nutritional security. Furthermore, our country gains foreign currency each year through the export of fish and fishery products.

The socioeconomic development of the nation depends heavily on fishing industry. The fisheries industry in Indian agriculture plays a significant role in the economic upliftment of the nation. It has gained recognition as a potent source of income and employment since it fosters the expansion of several subsidiary sectors and provides access to affordable, wholesome food in addition to earning foreign cash. Most significantly, it provides a significant portion of the country's economically underprivileged population with a means of subsistence. Nearly six million people are employed in the fisheries industry by India's resources, which include rivers, canals, ponds, lakes, and marine resources along the east and west coastlines of the Indian Ocean.

Capture Fisheries

Capture fisheries have been practised for thousands of years in both marine and freshwater habitats; however, the level of mechanisation and modernization of fishing in inland areas is generally inferior in many ways. India's marine capture fishery production potential is estimated to be 2.02 million Sq. km. with the Exclusive Economic Zone (EEZ) spread across the 8129 km of shoreline along the mainland and islands. The potential productivity of India's marine capture fisheries is projected to reach 3.94 MMT. However, because the majority of Indian marine capture production is restricted to shallower water areas, resources are being overexploited. The majority of fishing operations are focused on inshore waters using mechanised, traditional, and artisanal methods. In the Indian Exclusive Economic Zone (EEZ), there are abundant prospects for fishing in deeper waters. Millions of Indians participate in catch fisheries, and vast Indian resources have the potential to give millions more people access to jobs and business opportunities.

Fish production has large potential in India's interior fishing resources, which include 3.15 million ha of reservoirs, 1,91,024 km of rivers and canals, 2.35 million ha of ponds and tanks, 1.3 million ha of Oxbow lakes and abandoned waters, and 1.24 million ha of brackish waters. Despite the fact that fishing techniques in inland waters are less sophisticated and mechanised than those in the maritime sector, there are a lot of job and business prospects

Fishing Gears

The culture and capture of fish and shellfishes, and other aquatic species are among the many fishing activities that include the use of fishing nets. Over the centuries, fishing gears have progressed from being simple point-shaped wooden projections to sophisticated and complicated devices. The mall to large-scale industrial operations carry out the fabrication and manufacture of the nets, devices and other accessories. Diverse forms, styles, and quantities of nets and gears used in fishing activities are constantly in demand. A large portion of the Indian net manufacturing business exports high-quality fishing nets, traps, cages, and other fishing gears to most of the world's nations. A portion of the sector has developed expertise in the design and production of culture cages and other equipment's, which are in high demand globally. This industry can guarantee significant employment and earnings.

The different employment / entrepreneurship opportunities can be generated in this sector, which is mentioned below.

1. **Fishing Gear Innovation Incubators**: Create business incubators that provide mentorship and resources to help entrepreneurs create and test innovative fishing gear prototypes.
2. **Fishing Gear Rentals**: Provide fishing enthusiasts with a rental option for their gear so they can still enjoy the sport without having to invest in it.
3. **Fishing Gear Recycling**: Create strategies for recycling and reusing worn-out or damaged fishing gear, which will reduce waste and aid in environmental preservation.
4. **Fishing Gear Leasing**: Provide leasing services for fishing gear, so that fisherman can access new equipment's / devices without making a large upfront investment, encouraging the adoption of more effective instruments.
5. **Fishing Gear Exchanges**: Provide fishing equipment leasing services, so that fishermen may access modern gear without having to make a significant initial investment, promoting the adoption of more efficient tools.
6. **Eco-friendly Gear Manufacturing**: To lessen the industry's influence on the environment, concentrate on producing fishing gear and equipment with sustainable materials and manufacturing techniques.

7. **Fish Aggregating Devices (FADs)**: Create and use FADs to draw fish and increase catch rates, which will help local fishermen and provide a consistent supply for processing facilities.
8. **Fishing Safety Equipment Rentals**: To increase fishermen's safety, provide rentals for safety gear such life jackets, emergency beacons, and communication tools.
9. **Fishing Equipment Manufacturing**: Entrepreneurs can start companies that specialize in manufacturing high-quality fishing equipment, such as nets, lines, hooks, fishing rods, and other necessary tools. This helps the local fishing community as well as the larger fishing sec
10. **Customized Gear Solutions**: Create specialized fishing equipment that is suited to a particular fish species or fishing environment in order to increase productivity and decrease bycatch.
11. **Fishing Gear Innovation Competitions**: Organize contests to promote improvements in fishing gear, rewarding businesspeople with innovative ideas with rewards and providing support.
12. **Branding and Marketing**: Creating powerful brands for fishing equipment and seafood can provide businesses a competitive edge. Effective marketing techniques can highlight the products' quality, sustainability, and originality.
13. **Fishing Safety Solutions**: Create and promote safety gear, instruction, and procedures to increase the security of fisherman while they work.
14. **Ghost net Management:** Develop creative approaches to managing fisheries trash, such as ocean plastic cleaning programs, to solve the environmental issues brought on by discarded fishing gear.
15. **Fisheries Extension Services**: Organize consulting services that offer guidance to fishermen on how to increase productivity, reduce post-harvest losses, and employ cutting-edge fishing methods.
16. **Research and Development:** Innovations in fishing techniques, fishing gear, and materials can advance the industry. R&D can be funded by business owners that want to develop cutting-edge fishing technology.

Fishing Crafts

Fishing boats and other vessels are crucial, especially in the capture fisheries. A variety of fishing vessels, from small dugout canoes to enormous commercial fishing ships, are in high demand. Numerous jobs are created by the manufacture, building, and maintenance of fishing vessels. For routine operations, some fishing ships need hundreds of personnel. Additionally, routine maintenance, repair, navigation, harbouring, etc. provide a significant amount of job opportunities.

1. **Fisheries Aggregator Platforms**: Create websites and online platforms that compile resources for the fishing sector, including goods, services, crafts, gears and information.
2. **Fishing Vessel Services**: Offering maintenance, repair, and outfitting services for fishing boats can be a successful business. For their boats, many fisherman require routine upkeep and improvements.
3. **Cold Chain Infrastructure**: Maintaining the quality and freshness of seafood items is dependent on creating the necessary cold storage and transit facilities inside the vessel, particularly in a tropical nation like India.
4. **Export Facilitation**: Assist regional seafood exporters in navigating the difficulties of international trade laws, certifications, and market access.
5. **Technology Innovation**: The effectiveness of fishing operations can be greatly improved by technological advancements like GPS navigation systems, fish tracking devices, and fish detecting equipment. These technical solutions can be created and marketed by entrepreneurs.
6. **Sustainable Fishing Practices**: Entrepreneurs can concentrate on promoting ethical fishing methods to ensure long-term viability while safeguarding the ecosystem.
7. **Fishing App Development:** Create smartphone apps that give fisherman access to up-to-the-minute weather forecasts, market prices, fishing rules, and communication tools to increase their productivity and decision-making.
8. **Fishermen's Training Institutes**: Create specialized training facilities that provide in-depth instruction on fishing methods, safety precautions, business management, and sustainable practice
9. **Export and Processing**: Entrepreneurs can access the export market by setting up seafood processing and packaging facilities, taking use of the enormous marine resources of our nation.

10. **E-commerce Platforms**: Creating online markets for fishing gear, seafood goods, and services can effectively connect suppliers, fishermen, and customers.

11. **Training and Skill Development**: Establishing training facilities to instruct local fishermen in contemporary fishing methods, safety procedures, and ecological practices can empower them and develop a trained labor force.

12. **Eco-Tourism and Fishing Tours**: India's coastal regions are quite beautiful. Entrepreneurs can offer fishing trips, natural excursions, and cultural encounters that incorporate ecotourism and fishing.

13. **Fish Processing and Value Addition**: Installing fish processing equipment on the ship can help open doors for both domestic and international markets by turning raw seafood into products with added value including fillets, canned foods, fish oil, and fishmeal. This increases the catch's value and lowers post-harvest losses.

14. **Fisheries Insurance**: To reduce risks for fishermen and other industry stakeholders, offer specialist insurance packages for fishing vessels, gear, and aquaculture activities.

15. **Government Programs and Subsidies**: Business owners should be informed about government programs and subsidies that support the fishing industry like PMMSY (Pradhan Manthiri Matsya Sampada Yojana), such as funding for infrastructure construction, equipment purchases, and skill development. Awareness and education must also be taken towards the common people towards the governmental schemes making use of them. When pursuing prospects in the fishing craft and gear sector, business owners should take sustainability, local communities, and environmental regulations into account to ensure long-term success and a beneficial impact.

Other Entrepreneurship Options

1. **Supply Chain Management**: Entrepreneurs may focus on building efficient supply chains that connect fishermen with processing plants, wholesalers, and retailers. Ensuring a smooth transition of seafood products from the producer to the final consumer is essential for the industry's growth.

2. **Export and Market Intelligence**: Entrepreneurs may focus on export and market information, helping exporters of fish and seafood to find

international markets, client preferences, and legal needs.Advocate for policies that support the fishing industry, address regulatory issues, and advance sustainable practices by lobbying and advocating to government officials.

3. **Education and Awareness**: The creation of educational programs and awareness campaigns on sustainable practices, marine conservation, and ethical fishing can help the sector and the environment.
4. **Collaboration with NGOs**: Collaborations with non-governmental organizations focused on community development and marine conservation can help entrepreneurship's social and environmental aspects.
5. **Fish Waste Utilization**: Convert fish byproducts and trash into useful resources like fish oil, fish meal, or even items for the cosmetics and pharmaceutical industries.
6. **Climate-Resilient Solutions**: Create cutting-edge, climate-resilient fishing techniques and equipment to meet the demands of climate change and ensure the industry's long-term viability.These business options demand meticulous preparation, in-depth knowledge of the fishing sector, and dedication to sustainability, moral behavior, and community improvement. Entrepreneurs that can successfully negotiate these issues will be well-positioned to support the expansion and sustainability of India's fishing equipment and craft industry.
7. **Marine Environmental Monitoring**: Create technologies to track and evaluate how fishing operations affect the environment, promoting sustainable fishing methods and following rules.
8. **Fishing Knowledge Platforms**: Create websites or mobile applications that offer lessons, best practices, and educational materials to help fishermen develop their abilities and expertise.
9. **Fishermen Cooperatives**: Encourage the establishment of cooperatives so that fishermen may work together to obtain resources, bargain for better prices, and generally improve their standard of living.
10. **Seafood Retail Innovation**: Create innovative seafood retail ideas, such as themed seafood eateries or mobile seafood trucks, to meet the increased demand from customers for a variety of sustainable seafood selections.

11. **Fishery Product Branding**: Create a strong brand identity for seafood that is fished nearby, emphasizing its freshness, sustainability, and local ties.
12. **Fishermen Training Centers**: Create facilities where seasoned anglers can instruct newbies in both conventional and contemporary fishing methods, passing on invaluable skills and expertise.
13. **Fisheries Waste Recycling**: Create creative recycling techniques that can transform waste from fishing into beneficial items like compost, biofuels, or biodegradable plastics.
14. **Fishing Research and Development**: Make an investment in R&D to create innovative fishing techniques, machinery designs, and sustainable business practices.
15. **Fish Market Analytics**: Make informed decisions for fishermen and processors by using data analytics to analyze market trends, price swings, and demand patterns.
16. **Fisheries Investment Groups**: Create investment groups or funds that are dedicated to funding sustainable fishing projects, helping the industry flourish.

Government Initatives to Fishing Gear and Craft Entrepreunership

1. **Promotion of deep-sea fishing:** The government of India offers 60% monetary assistance to the traditional fisherman to convert their vessels to deep sea fishing vessels. Deep sea fishing, which is done beyond the EEZ of 200 nautical miles from the shoreline, will be more appropriate to fulfill the upcoming surplus demand of food.

Fig. 13.3: Large mesh purse seine operating off Cochin (*Courtesy*: CIFT, Cochin)

2. **Submarine tourism:** The state of Gujarat is planning to launch India's first Underwater sub-marine tourism in Gujarat state for the tourists to explore the sunken land of Krishna Dwaraka. The government aims to bring in the project to action during 2024. The submarine is planned to be designed to develop up to 100 meters underwater encompassing 24 passengers within it. The Gujarat government has signed an MOU with the Mazagon Dock Shipbuilders Limited for the construction of submarines.

Fig. 13.4: Graphical representation of the proposed submarine tourism in Gujarat (*Courtesy*: India Today)

3. **Glass bottom boats:** The Gulf of Mannar Marine Biosphere Reserve Trust (GoMBRT) is working on to boost the launch of glass bottom boats to promote the ecotourism in the area. Glass bottom boats has been purchased for this scheme, where tourists can view corals and underwater biodiversity from the glass bottom boat with 20 members additionally having snorkeling facilities also. 60% of the revenue generated will be spent for conservation of corals and rest 40% will be spend towards the development and upliftment of the coastal community. Glass bottom boats initiatives are already in on at Andaman and Lakshadweep islands.

Fig. 13.5: Tourists travelling in glass bottom boats in Gulf of Mannar
Courtesy: S. Godson Wisely Dass

4. **By Catch Reduction Devices and Turtle excluder device**: By-catch reduction devices had been developed by CIFT and extensive field operations and demonstrations were done. The existing turtle excluder device design has been modified by ICAR - CIFT (Central Institute of Fisheries Technology) based on the recommendation of Marine Products Export Development Authority (MPEDA) and National Oceanic and Atmospheric Administration (NOAA US based scientific regulatory agency within the US department of commerce) as well as considering the needs of the Indian trawling sector. Trial launch were done late in 2023 and recommendation to install this model of TED in trawlers are advised by the Government.

BRD- Squaremesh in Belly portion of the net

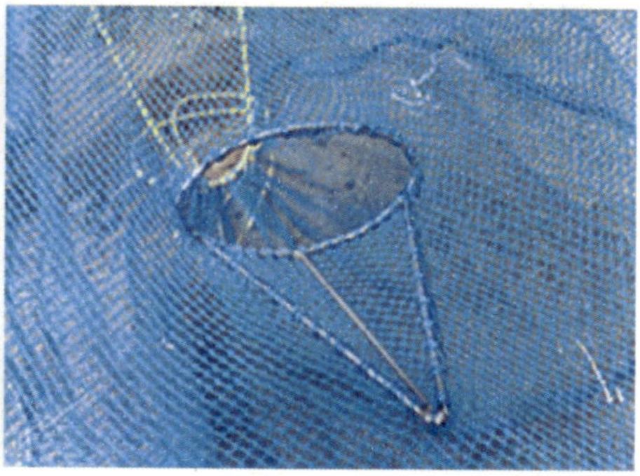

BRD- Conical Bycatch reduction device

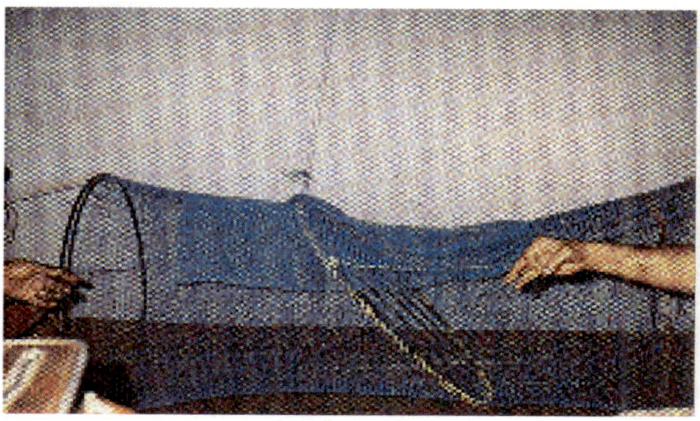

TED- Being fitted behind the cod end

Escaped turtle caught in the Collecting net

Fig. 13.6: By-catch reduction devices and Turtle excluder Device
Courtesy: MSSRF website

Fig. 13.7: Field trials of Turtle Excluder Device by MSSRF, MPEDA and ICAR-CIFT

Conclusion

Successful entrepreneurship in the fishing craft and gear sector requires adaptability, a commitment to sustainability, and a thorough understanding of the local fishing industry. By focusing on innovation, community engagement, and responsible practices, entrepreneurs can contribute to the growth and development of this vital sector in India. Strong understanding of the industry, regulations, environmental considerations, and the needs of the local communities and fishermen. Entrepreneurs who approach this sector with innovation, sustainability, and community development in mind are likely to find meaningful and profitable opportunities. Entrepreneurs that value innovation, sustainability, and community involvement can have a huge impact while creating profitable businesses as the fishing craft and gear market continues to develop. To be competitive and relevant, it is essential to keep up with market trends, technology developments, and shifting consumer tastes.

References

Annual Report, 2017-18. Department of Animal Husbandry, Dairying and Fisheries, Ministry of Agriculture and Farmer Welfare, Govt of India.

Annual Report, 2016-17. Department of Animal Husbandry, Dairying & Fisheries Ministry of Agriculture and Farmer Welfare, Govt of India.

Ayyappan S. and A.D. Diwan, 2006. Fisheries Research and Development in India, Fishing Chimes, 26(1).

FAO. 2022. The State of World Fisheries and Aquaculture 2022 - Meeting the sustainable development goals. Rome.

Handbook on Fishery Statistics, 2022. Department of Animal Husbandry, Dairying and Fisheries, Govt of India.

Markad, S. S., Gaikwad, J. M., Patil, D. W., & Shinde, V. D. (2019) International Journal of Management Economics 1 (29) Review Of Research.

Murthy, L. N., & Ninan, G. (2021). Fishpreneurship opportunities. ICAR-Central Institute of Fisheries Technology.

National Fisheries Development Board {NFDB}, Department of Animal Husbandry, Dairying & Fisheries, Ministry of Agriculture and Farmers Welfare, Government of India

Open AI, ChatGPT (2023).

Sipra Karmakar (2023) Trend and Pattern of Indian Seafood Export, Journal of Survey in Fisheries Sciences 2023, 10(1S) 4995-4999

https://www.mssrf.org/small-news/mssrf-partners-for-turtle-excluder-device-field-trials/

https://www.newindianexpress.com/states/tamil-nadu/2022/mar/23/glass-bottom-boats-to-promote-eco-tourism-in-thoothukudi-2433296.html

https://www.news18.com/india/gujarat-govt-signs-mou-with-mazagon-dock-shipbuilders-for-tourist-submarines-to-dwarka-8720856.html

https://www.indiatoday.in/india/story/gujarat-to-launch-countrys-first-submarine-tourism-in-lost-city-dwarka-lord-krishna-2481709-2023-12-29

14

Peri-urban Aquaculture as an Opportunity for Entrepreneurship in India

P. Routray*, D.K. Verma, K. C Das and N.K. Barik

ICAR-Central Institute of Freshwater Aquaculture, Kausalyaganga Bhubaneswar 751002, Odisha

Abstract

This chapter gives an overview of peri-urban aquaculture using Biofloc and such new technologies and entrepreneurship opportunities in these fields.

Keywords: *Aquaculture, Peri-urban aquaculture, Biofloc technology*

Introduction

Aquaculture has emerged as a viable business model having high growth potential for the entrepreneurs and start-ups in India. The rate of adoption of the aqua-business models for new age entrepreneurs has been very high in recent times. Some of the viable business models that are technology based and scalable have been attractive to the entrepreneurs are in the indoor aquaculture system like RAS, Biofloc systems. It offers many advantages like low space requirements, easy monitoring, higher level of intensification, easy management, short supply chain to the urban markets etc. These indoor aquaculture system are broadly defined as peri-urban aquaculture system offering high growth potential to the entreprenuers in India. The chapter provides a broad outline of the opportunities and recent trends of entrepreneurial developments through peri-urban aquaculture.

Entrepreneurship is the risk taking ability and readiness to venture into a new economic enterprise, along with all its challenges in order to make profit. One of the examples of entrepreneurship is the starting of new businesses in any field with business potential. In other words, entrepreneurship is the creation or extraction of economic value (Diochon et al. 2011; Gaddefors et al.

2017; Alvarez, et al. 2001). It is also viewed as a change, generally entailing risk beyond what is normally encountered in starting a business, which may include other values than simply economic ones. A person gets the benefits as he invests in one or more businesses, bearing most of the risks and availing most of the rewards. According to Shane (2000), entrepreneurship comprises both "enterprising individuals" and "entrepreneurial opportunities"[4]. An entrepreneur is generally seen as an innovator, a source of innovative ideas, services, and business.

Entrepreneurship in aquaculture

Every economic activity has an underlying entrepreneurship opportunity that needs to be identified by the person who successfully makes that into an enterprise. There are several areas of entrepreneurship opportunities in agriculture and aquaculture in India. The entrepreneurship opportunities vary in scale scale based on the market perception, which includes demand for a particular product. In aquaculture, there are several traditional farming practices available where entrepreneurship opportunities exists viz. fish seed hatchery, seed rearing, seed and fish transport, seed packaging, fish processing, ready to cook fish products and nets and gears. Apart from this many new opportunities are being created or available for adoption by entrepreneurs. Some of these opportunities are circulatory aquaculture, feed additives, feed making, and black soldier fly production. One of the most promising areas that has recently drawn attention is the periurban aquaculture using medium and large tanks made up of geomembrane fabrics with round the clock aeration and monitoring facilities. In this periurban aquaculture practice, biofloc method is employed to reduce water usage and economize on feed. Besides this one of the major advantages of this is that it can be practiced in urban and periurban areas where land holdings are very less.

The consumer demand for fish and fisheries products is increasing steadily due to population growth and changes in consumption patterns facilitated by a rise in per capita income and gradual diversification of dietary habits. It is further facilitated by the fact that consumers want fresh and healthy fish at ease near to their home. There is a need to enhance fish production and productivity through development of successful entrepreneurs in the sector as the demand for fish and fisheries products is growing in domestic and global markets. We have to showcase the opportunities available in the sector such as periurban aquaculture to the enthusiastic youth of our country. By doing this, we can foster entrepreneurship in fisheries and aquaculture sector. Once a confidence sets in the system then the capital influx and innovation become a part of the ecosystem.

Aquaculture is defined by Food and Agriculture Organization (FAO), as "farming of aquatic organisms including fish, molluscs, crustaceans and aquatic plants. Farming implies some form of intervention in the rearing process to enhance production, such as regular stocking, feeding, protection from predators, etc. Farming also implies individual or corporate ownership of the stock being cultivated. Application of science and innovation is realized in terms of increased income, employment and consumption. Global food fish consumption increased at an average annual rate of 3.1 percent from 1961 to 2017, a rate almost twice that of annual world population growth (1.6 percent) for the same period, and higher than that of all other animal protein foods (meat, dairy, milk, etc.), which increased by 2.1 percent per year. Per capita food fish consumption grew from 9.0 kg (live weight equivalent) in 1961 to 20.5 kg in 2018, by about 1.5 percent per year (FAO, 2020). The four most primary resources of the freshwater aquaculture developments are land, water, human capital and inputs. At the beginning of the aquaculture development, water based resources like wetlands, impoundments, depression, natural water bodies were used for fish farming. In later parts, water bodies created out of the land-based resources like paddy fields, swampy lands, water logged lands etc. to create aquaculture production environment. Due to the pressure on these resources and for economic reasons, It is also expected that the urban lands like terraces, indoor sheds, home back yards will increasingly be used for creating space for the aquaculture activities in future. Therefore, future aquaculture expansion is moving towards land and indoor based activities with more sophistication and less water use. Social and economic context of aquaculture development has undergone considerable changes in recent times. Consumers of the fishes demand best quality fish that are produced in healthy environment free from pollutants and disease. There is increased awareness on the quality of environment in which fish is grown, traceability, freshness etc. Producers of the fish are looking for an environment in which fish production can be intensified. With global climate change, urbanization, completion for resources, land and water are becoming increasingly scarce. Producers of fish are looking for land and water saving technologies. In the traditional aquaculture system, extensive aquaculture was practiced which involved low intensive cultivation with use of relatively higher quantities of land and water. These trends are reversed and farmers are adopting more intensive system of cultivation which reduces quantities of land and water use.

New Business Models

Periurban aquaculture involving indoor production of aquaculture has emerged as new enterprises for new age entrepreneurs. Backward linkages with production of inputs like seed and feed and forward linkages with marketing, processing, value addition are proved to be attractive option of the start-ups and entrepreneurs. These are new business entrepreneurship models developed as to cater to the need of the urban consumers. Several new advancements in water filtration, purification and recirculation make it possible for indoor and backyard fish farms to grow in size and production capacity. A regular fresh and large quantity of the fish can be produced and consumed within the urban environment involving lower cost of handling and marketing. Within these premises, large numbers of entrepreneurs are exploring business models suitable to the markets. Similarly, vertical fish farming is also gaining importance as this offers a great scope for increasing food production in a limited space. In the new business models, the advantages offered by the urban environment especially on proximity to the market, compactness of the production system, control of the production process etc. are made use of. The resources used are in the form of the seed, feed, water, etc. and output are in the form of diverse products of fish are typical example of the peri-urban business model.

Concept of Periurban Aquaculture

Concept of periurban aquaculture is in vogue for many years as farmers are developing urban based agri-aquaculture system with limited use of land and water. Such systems require migrating from low intensive to high intensive aquaculture techniques. Such systems are placed close to local communities and provide steady supply of fresh fish to the nearby market at affordable price. There are encouragements for adoption of such technologies by government policies. New policies on the starts up, entrepreneurship development with the objective to create employment through adoption of modern technologies are providing new incentives for new age aquaculture. Business incubation centres are providing handholding support for such aquaculture. Now, many entrepreneurs are attracted towards developing periurban aquaculture systems in their small land holdings and even in their kitchen gardens. Some of the components of the periurban business models are as below:

- Establishment of indoor production facilities (Tank based aquaculture)
- Production and linkage with forward and backward enterprises
- Storage and transportation of fish

- Processing and value addition
- Branding and marketing of locally produced fish

Benefits of Periurban Aquaculture (PA)

The main benefit of periurban aquaculture is that it takes up less space than traditional aquaculture. It is ideal for small-scale farmers and those operating in limited spaces, such as urban and semi- urban areas. When periurban aquaculture is combined with biofloc technology and RAS it allows for more efficient use of water and feed resources. Periurban aquaculture is also beneficial because the fish are raised in a round the clock monitored environment, the risk of disease and its transmission is less compared to open water systems. This makes it a more sustainable form of aquaculture. Many horticulture/ garden crops can be grown year-round near the PA facility that will utilize the nutrient rich water from these systems. PA is an attractive option for locations with limited growing space or climates that are not conducive to agriculture. In recent years, the idea of urban fish farming has gained popularity as a way to provide food and income for city residents. Moreover, many people have become more interested in eating locally produced, fresh and sustainable food. The opportunities for city dwellers are many as potential sites for aquaculture in cities includes rooftops, parking garages, open space gardens and even inside buildings.

Limitations of Periurban Aquaculture

- There are also several limitations to Periurban Aquaculture viz. the technology is still relatively new and has not been tested on a large scale.
- The cost of setting up and running a PA system is sometimes prohibitive for some farmers.
- Few types of fish can be raised on a PA, it is limited to those that are able to thrive in close quarters and complete feed based systems.
- Tilapia and Pagasius and few other warm-water fish are ideal candidates for PA, but production cost is increasing in recent days due to the increasing feed cost.
- Sometimes slight lapse in monitoring and disease outbreak can lead to mass mortality.

Technological innovations in Periurban Aquaculture

New technological innovation like Re-circulatory system, biofloc system, aquapoics, hydroponics etc are some of the examples of the technological progress which have contributed to the development of new business models and has a great potential for entrepreneurship development in periurban areas. In the recent times, there are many examples of application of such technologies in practice. Technological development requires a long term process of development, standardization, packaging etc. However, upon standardization and pilot testing, it is easy for the farmers to adopt the technologies in large numbers. Generally periurban aquaculture involves a sandwich technology like minimal exchange water and low density biofloc systems (BFT). However, more intensive BFTs are also being practiced in periurban areas opening new vistas in aquaculture production.

Use of Biofloc in Periurban Aquaculture

Biofloc is a consortium of particulate matter formed predominantly by a biota of aerobic and heterotrophic bacteria, protozoa, microalgae (diatom), metazoans, detritus, faeces, remnants of feeds and remains of dead organisms.

Major requirement of Biofloc Technology in PA

- High stocking density
- High aeration (to maintain dissolved oxygen of >6 ppm)
- Poly lined ponds/ Poly tarpaulin ponds/ Cemented tanks
- Sludge/ waste removal system
- Indoor/ outdoor tanks
- Suitable candidate species
- Zero water exchange
- Zero tolerance to antibiotics

Composition and Nutritional Value of Biofloc

- The dry weight protein ranges from 25 – 50%
- Fat ranges 0.5 – 15%
- It is a good source of vitamins and minerals, particularly phosphorous
- It has an effect similar to probiotics

- The dried biofloc is proposed as an ingredient to replace the fishmeal or soybean in the feed

Advantage of BFT

- Eco-friendly culture system.
- It reduces environmental impact.
- Judicial use of land and water
- Limited or zero water exchange system
- Higher productivity (It enhances survival rate, growth performance, better feed conversion in the culture systems of fish).
- Higher biosecurity.
- Reduces water pollution and mitigate the risk of introduction and spread of pathogens
- It reduces utilization of protein rich feed and cost of standard feed.
- It reduces the pressure on capture fisheries i.e., use of cheaper food fish and trash fish for fish feed formulation.

Species suitable for Biofloc Culture in PA

Biofloc system is most suitable for species that can tolerate high solids concentration in water and are generally tolerant of poor water quality. Some of the species that are suitable for BFT are:

- Air breathing fish like Singhi (*Heteropneustes fossilis*), Magur (*Clarias batrachus*), Pabda (*Ompok pabda*), Anabas/Koi (*Anabas testudineus*), Pangasius (*Pangasianodan hypophthalmus*)
- Non air-breathing fishes like Common Carp (*Cyprinus carpio*), Rohu (*Labeo rohita*), Tilapia (*Oreochromis niloticus*), Milkfish (*Chanos chanos*)
- Shellfishes like Vannamei (*Litopenaeus vannamei*) and Tiger Shrimp (*Penaeus monodon*)

Table 14.1: Details of PA using Biofloc System

Sl.No	Component	Details
1	Area for 7 tanks	200 m2
2	Biofloc Tank size	4 metre diameter and 1.5 meter height (1.20 m water depth)

3	Water holding capacity of each tank	15,000 Litres capacity
4	Water quality parameters	Dissolved Oxygen-5mg/L, Temparature-26-34°C, pH-7.5 to 8, TDS-600ppm, Floc density-25-40 mg/l, Ammonia-0.5 ppm, Nitrite-0.3 ppm, Nitrate-150 ppm, Alkalinity-120-280 ppm
5	Stocking density	100 nos/m3 (1000 no.s per 15,000 litres tank depending on species)
6	Species cultured	GIFT Tilapia (*Oreochromis niloticus*)
7	Survival (%)	80
8	Type of feed to be used	floating pellet feed, @ 3% BW
9	Feeding frequency	4 times early stage, later 2 times per day
10	Duration of culture	6 months
11	Size/weight of the species(gm)	500 gm average weight
12	No. of crops per year	2
13	Production	4.2 Tonnes per crop (600kg per tank per crop)
14	Farm gate price(Rs)	130/- kg fish
15	Capital cost	6.00 Lakhs
16	Input cost	1.5 lakhs per one crop
17	Total project cost	7.5 lakh

Table 14.2: Cost Estimate of PA using Biofloc Systems with 7 tanks

Sl.No.	Components	Numbers	Cost (Rs.)	Total (Rs. In lakhs)
Capital Cost				
1	Setup of Tarpaulin/ Fibre tanks (15,000 Litres capacity)	7	25000	1.75
2	Shed material and accessories fixing charges	200 m2	120000	1.20
3	Water supply borewell (3 HP)	1	100000	1.00
4	PVC Pipe fittings for air, water flow	LS	75000	0.75
5	Nets and accessories	5	3000	0.15
6	One Blower (1 HP), Air stones and other accessories	1	30000	0.30
7	Electrification	LS	10000	0.10
8	Power Generator (2 KVA)	1	45000	0.45
9	Weighing balance	1	5000	0.10
10	Miscellaneous expenses			0.20
Total Capital Cost				**6.00**
***Input Cost for 1 Crop**				
11	Seed cost, Feed cost, Probiotics, Test kits etc.			1.50
Total input for 1 crop				**1.50**
Grand Total				**7.50**

*Input cost may vary depending on stocking density

Table 14.3: Economic feasibility (one crop) from 7 tanks

Sl.No	Component	Details
1	Capital Cost	6.00
2	Operational Cost	1.50
3	Total Project Cost	7.50
4	Gross Income per crop	5.46
5	Gross income at the end of one crop after deducting the recurring cost for the 2nd crop	3.96
6	Gross income from the 2nd crop	5.46
7	Gross income at the end of 2nd crop	9.42
8	Depreciation/maintenance @ 15% of capital cost	0.975
9	Interest @ 12% of TPC	0.90
10	Repayment @ 1/7th of TPC	1.07
11	Recurring cost for the next crop	1.50
12	Net Profit at the end of 2nd crop 9.42÷ (0.975+0.9+1.07+1.50)	4.975

Species: Monosex Male Tilapia (*Oreochromis niloticus*)

Table 14.4: Technical details of biofloc system

Items	Unit	Quantity
No. of Cultured Tank	units	10
Capacity of each Cemented Tank	litres	1,200-1500
Stocking	numbers	40
Stocking size	grams	20-30
Culture Period	days	120-150
Final Harvesting	grams	200-300
Final Yield	kilograms	9-10.5
Feed Consumption	kgs	12-15
Feeding rate	of body weight/day	2-3%
Water Exchange	per week	20-30%
Total Water exchange	litres	7000-8000

Table 14.5: Types of Tanks (Costing)

	FRP made CT	HDP	Cement Concrete Tank	Low Cost Cement Tank
Price	24,000	12,000	12,000	5,000
Longitivity (years)	20	10	30	10
No. of Culture	60	30	90	30
Cost per Culture	400	400	133	**167**

Production of Tilapia and IMC Fry in PA System using BFT

To assess the growth of 17 α methyl testosterone (MT) treated Nile tilapia, *Oreochromis niloticus*, a total number of 500 tilapia with average weight 49 g were stocked in two 7m^3 tanks under biofloc system and one tank without biofloc (control). The stocking density was 100 numbers/cubic meter water. The sampling of MT treated tilapia in cemented tank biofloc culture after six months (August 2020 - January 2021) showed an average growth of 302 grams and average length of 310 mm under bioflock system and 241g in control tank. The floc density was maintained at 20-25 ml as measured by Imhoff cone. The total survival rate was 98% and 99% in biofloc and control tanks respectively. A final harvest of 156.8 Kg and 119.2 Kg were recorded from biofloc and control tanks respectively. The final economics of the culture is shown in the table - 1.

Table 14.6: Economics of tilapia culture in PA using biofloc tanks of 7 m3 capacity (for one tank)

Items	**Biofloc tank**	**Control (Non-biofloc tank)**
No of fish stocked	500	500
Average size stocked	49 g	49 g
Final yield (Kg)	156.8	119.2
Final survival rate (%)	98	99
Feed given (Kg)	182	181
Weight of carbon source (jaggery and rice water)	10	NA
Cost of seed (Rs.)	1500	1500
Cost of feed (Rs.)	7280	7240
Cost of Carbon source (Jaggery/rice water) (Rs.)	1000	0
Electricity (aeration) (Rs.)	900	900
Other operational cost (Rs.)	1000	1000
Total Cost (Rs.)	11680	10640
Selling price at farm gate (Rs.)	100	100
Gross sale	18200	11920
Profit	6520	1240
Profit %	35.8%	10.40
Cost of Production (Rs.)/Kg	74.48	89.59
FCR (without jaggery/rice water)	1.37	1.91
FCR (including jaggery/rice water)	1.45	NA

Performance evaluations of Indian major carp fry production

To assess the survival and growth of spawn (yolk sac absorbed larvae) of Indian major carps in biofloc system a trial was conducted using 20000 litre capacity tanks. A stocking density of 15000 spawn /m^3 followed in three tanks. The spawn grew to 30-40 mm in twenty days in a low floc density of 5-10 ml/1000 ml. A survival rate of 40-50% was achieved from biofloc tanks. No water exchange was made during this period and a powdered feed containing 30% CP was provided ad lib to the reared spawn. The important take home message from this study was that fry production of IMC can be undertaken in biofloc systems that will reduce dependency on the nursery fry production in earthen ponds in future. This has the potential to reduce the drudgery involved in the nursery preparation and environmental uncertainties like flooding etc. Similar experiments with fry (20-30mm) rearing for production of fingerlings showed promising results.

Table 14.7: Economics of IMC fry (rohu and mrigal) production in 20 K capacity biofloc tanks (for one tank)

Items	Biofloc tank	Control (Non-biofloc tank)
Water holding	15000 Litres	15000 Litres
No of spawn	225000	225000
Average size stocked	Yolk sac absorbed	Yolk sac absorbed
Final siz (mm)	30-40	30-40
Final survival rate (%)	45	25
Feed given (Kg)	3	4
Weight of carbon source (jaggery and rice water)	10	NA
Cost of seed (Rs.)	900	900
Cost of feed (Rs.)	210	280
Cost of Carbon source (Jaggery/rice water) (Rs.)	100	0
Electricity (aeration) (Rs.) 20 Units	500	500
Other operational cost (Rs.)	2500	2500
Total Cost (Rs.)	4180	3710
Total Production of fry	67500	37500
Selling price of fry (Rs./1000)	300	300
Gross sale	20250	11250
Profit	16070	7540
Cost of Production (Rs.)/fry	0.06	0.09

Conclusion

The periurban aquaculture would definitely help in income generation for entrepreneurs including women those who are only looking after the domestic works. Moreover, the unutilized and underutilized areas will be put to use for economic benefits in periurban areas using these technologies.

Refrences

Alvarez, Sharon A.; Busenitz, Lowell W. (1 December 2001). "The entrepreneurship of resource-based theory". Journal of Management. 27 (6): 755–775. doi:10.1177/014920630102700609. ISSN 0149-2063. S2CID 220587830.

Diochon, Monica; Anderson, Alistair R. (1 March 2011). "Ambivalence and ambiguity in social enterprise; narratives about values in reconciling purpose and practices". International Entrepreneurship and Management Journal. 7 (1): 93–109. doi:10.1007/s11365-010-0161-0. hdl:10059/613. ISSN 1555-1938. S2CID 144081539.

FAO. 2020. The State of Food and Agriculture 2020. Overcoming water challenges in agriculture. Rome. https://doi.org/10.4060/cb1447en

Gaddefors, Johan; Anderson, Alistair R. (1 January 2017). "Entrepreneurship and context: when entrepreneurship is greater than entrepreneurs". International Journal of Entrepreneurial Behavior & Research. 23 (2): 267–278. doi:10.1108/IJEBR-01-2016-0040. hdl:10059/2299. ISSN 1355-2554.

Shane, S. (2000). "The Promise of Entrepreneurship as a Field of Research". Academy of Management Review. 25 (1): 217–226. doi:10.5465/AMR.2000.2791611.

15

Entrepreneurship Opportunities in Coldwater Fisheries in India

Atul K. Singh

Former Director ICAR-Directorate of Coldwater Fisheries Research & Ex-Emeritus Scientist, ICAR-National Bureau of Fish Genetic Resources Lucknow-226002 (UP), India

Abstract

Coldwater fish species particularly introduced non-native trout, and endemic mahseer species are highly valued for their taste and nutritional qualities, and are often sold as premium products. Advances in technology, which includes recirculating aquaculture systems (RAS) and closed containment systems, are replacing the traditional farming of coldwater species. New systems allow for better control of water quality and temperature, and can reduce the risk of disease and pollution. Coldwater aquaculture generally does not have any negative environmental impacts, however, intensification, thereby release of excess nutrients, and metabolites, use of antibiotics, and pollutants, can harm wild fish populations and their habitats. Coldwater aquaculture industry in the Indian Himalayan region (IHR) is an important part of the fishery sector, with a focus on producing high-value species like trout and mahseer for domestic and export markets and also for recreational fisheries. Coldwater fishery industry is evolving fast, with incorporation of new technologies and sustainable management practices creating nutritional security as well as entrepreneurship opportunities.

Keywords: *Coldwater, aquaculture; trout; mahseer; rearing technology*

Introduction

The Himalayan and peninsular regions of India comprise of diverse coldwater resources in terms of upland rivers, fast flowing streams, high, mid and low altitude natural lakes and reservoirs (Singh and Akhtar, 2015; Singh and Sarma, 2017; Sarma et al, 2018). The Indus, the Brahmaputra and the Ganga basins in the Indian Himalayan Region (IHR) is a domicile of over 4% of the country's population having rich repository of biodiversity, a natural habitat of over

8000 species of flowering plants, 816 tree species, 675 edibles, nearly 1740 species of medicinal plants and about 258 species of fish (Singh and Akhtar, 2015; Pramanik and Bhaduri, 2016; Singh and Sarma, 2017). IHR mainly encompasses the states of Jammu & Kashmir, Himachal Pradesh, Uttarakhand, Sikkim and Arunachal Pradesh where the diverse natural resource-base and wide climatic range, harbour huge inhabitants of endemic, indigenous and exotic coldwater fish species extending an enormous scope of developing upland farming practices (Singh and Sarma, 2017; Singh, 2020), game and ornamental fisheries of large extent (Zaidi et al., 2018). These can provide an excellent opportunities for augmenting livelihood security of upland people residing in the Himalayan terrain, where opportunities to earn revenue is difficult and hard to adopt the farming practices. Today, the Himalayan states offer a unique value proposition in coldwater fisheries to harness its potential. There is an increasing trend of coldwater fish production from the present level of 52,084 MT to 90 thousand MT by FY 2024-25 boosting current productivity from ~1 ton/ha to 3 ton/ha. Coldwater fisheries in the country is now occupying a key place among India's freshwater fishes, contributing significantly to the nutritional security and job creation in rural hilly areas. More than 1.40 lakh fishermen in India directly depend on coldwater fisheries in hilly areas. In Himalayan region, aquaculture and fisheries represent an essential, often irreplaceable source of high quality animal protein crucial to the balance of diets in marginally food secure communities. The aquatic resources in the IHR are quite valuable for the development of aquaculture and fishery for food, sport, recreation and employment. In this paper, the author has attempted to present possible opportunities in coldwater aquaculture and fisheries which has great promises of future for job creation and nutritional security.

Coldwater fishery resources

The coldwater fishery resources in the IHR are spread from central, north-western to north-eastern Himalaya, and some parts of Western Ghats which are presented in table 1.

Table 15.1: Coldwater fishery resources of Indian Himalayan region, India

State	Rivers & canals (thousand km)	Water bodies (lakh hectare)
Jammu & Kashmir	27.78	0.30
Himachal Pradesh	3.00	0.43
Uttarakhand	2.69	0.20
Sikkim	0.90	0.03
Arunachal Pradesh	2.00	0.04

Two fundamental types of fisheries are existing in coldwater fisheries sector- (i): subsistence fishery (ii) sport/recreational fishery. The coldwater fisheries resource harbours 258 species belonging to 21 families and 76 genera. Out of that 203 species are found in Himalaya and 91 species are found in Deccan plateau (Singh and Akhtar, 2015). Generally commercial fish production in mountain area is low or practised on a very limited scale. Low biological productivity usually results in the occurrence of small-sized fish in river streams. Water temperature is always the limiting factor in fish production. The subsistence and commercial fisheries exploit carps (*Labeo* and *Tor* sp.), schizothoracines (*Schizothorax* and *Schizothoraichthys* sp.), garrids (*Garra* sp.), lesser barils (*Barilius* sp.), and sisorids (*Glyptothorax* and *Glyptosternum* sp.). The other genera are of small size and of low economic value (Singh and Akhtar, 2015; Sarma et al., 2018). The introduced non-native rainbow trout (*Oncorhyncus mykiss*) and brown trout (*Salmo trutta*) are important commercial aquaculture species of the Himalaya especially in the high altitudinal region (Singh, 2020; Singh and Srivastava, 2023) for food and recreation. Mahseer the king of game fish (*Tor* & *Neolissochilus*) provides valuable resources for development of sport fisheries and eco-tourism in the country. Mahseer and introduced non-native trout particularly brown trout available in coldwater regions are highly demanded among the anglers (Singh and Srivastava, 2023). The IHR is an abode for sport fishery where anglers from all over the world flock to the Indian rivers to try and hook the mighty mahseer or trout. About 3,800 km lengths of river and stream stretches in the upland region of our country are holding sizeable mahseer and 714 km stretches for brown trout, for angling purpose .

Fish production status

Though coldwater fish production through aquaculture practices is gaining momentum in the recent past, however maximum share comes from capture fisheries. At present the total fish production from coldwater sector contributes approximately 2% of total inland fish production. It is important to mention that aquaculture production potential of the coldwater sector has not been utilized to its maximum. Therefore, to enhance the fish production in IHR states, gifted with natural lakes and reservoirs could be better utilized under culture based capture fisheries program utilising modern culture tools and practices. Initiatives for open water cage culture in Himalayan lakes and reservoir could also be taken up more intensively for the stock enhancement of mahseer and trout besides carp (Sarma et al., 2018; Singh, 2020). Traditional fish farming is not in regular practice in most of the hill states. Therefore, there is immense scope for the development of aquaculture utilising modern methods to provide a source of employment to the people of upland region. Many suitable sites are available in different parts of the Himalayan region, which could be utilized for fish production. The suitable sites are lying along the banks of the rivers, rivulets, streams and streamlets. Depending upon the micro-climatic conditions of the region, such suitable patches of water bodies should be identified throughout the region and be brought under aquaculture practices. Today, the available technologies allow the culture of a number of exotic and indigenous coldwater fish species in Indian Himalaya. The most common species for coldwater aquaculture are rainbow trout, common carp, Chinese carp, minor carp along with mahseer and other endemic fish species.

(i) Trout production in India

Two species of trout the brown trout (*Salmo trutta*) and rainbow trout (*Oncorhynchus mykiss*) are cultured in India (Singh, 2020; Singh & Srivastava, 2023). India has significant coldwater aquatic resources spread over the Himalayan and peninsular region. The first attempt to introduce trout was from England to the state of Jammu and Kashmir dates back to 1899 for sport and food (Singh and Lakra, 2011). The trout farming is now a commercial venture in India as in other Asian countries, but the states of Jammu & Kashmir, Himachal Pradesh, Uttarakhand, Arunachal Pradesh and Sikkim are providing sizeable amount of trout in the Himalayan region (Singh, 2020). These species can withstand temperature variation of 0-20 °C, but spawning and growth occurs in a narrower range (12-18 °C). The culture practices involves spawning, taking quality eggs from healthy brood fish, incubation of eggs, rearing of young fries in nursery ponds, raising of fingerlings in growing ponds and producing yearlings in raceways, circular ponds etc. (Pandey & Singh 2018; Singh, 2020). It is assessed that presently over 2100 metric tons of trout are produced (Figure 15.1A. & B). However, there is huge potential to raise this overall production further significantly (Singh, 2020). Eyed ova seed of the genetically improved variety of rainbow trout was imported from Denmark and being reared at trout farms at Beerwa and other places in Jammu & Kashmir, Hamni and Batahar in Kullu, Himachal Pradesh, Barot Mandi, Thalla in Chamba district of Himachal Pradesh, Dhamwari farm in Shimla, Himachal and Sangla in Kinnaur, Himachal Pradesh and also in Uttarkashi in Uttarakhand for larger productions. The culture of improved strain of trout will help these states to double the present trout production level of 2100 metric tons and target a production of over 5000 metric tons in the next five years.

A B

Fig. 15.1: (A). Upward trend of rainbow trout production in India **(B)** A catch of rainbow trout *Oncorhynchus mykiss.*

There has been increasing recognition that trout aquaculture can contribute to blue and green growth and the bio-economy of the mountain people. The use of enormous resources in. the IHR for food security and public health by sustainable trout production of high quality is the need in the present time. In Jammu and Kashmir, the annual total cost and gross return per farm were estimated to be Rs. 2.36 lakh per farm and Rs. 4.30 lakh per raceway, with a benefit cost ratio of 1.83, indicating the business economic viability. In Sikkim, a trout raceway of 51 m^3, which costs around 1.20 lakhs, has been observed to produce 500 kg of trout fish at an average price of 800 per kg, yielding an income of 4 lakhs. The net profit from a single raceway would come to ₹ 2.7 lakhs, which is a very venturing and reasonable profit to the farmers. Trout farmers are usually desperate for investing substantially on feed per kg fish production. Therefore, the author has developed feeding schedule for trout to maximise their production at different temperature regimes in the IHR optimising quantity of the existing feed available with state governments for better FCR. This new feeding schedule will not only optimise the use of feed but also curb production of ammonia, and other harmful products in the discharged water regulating environmental deterrents besides raising the biomass on economic terms (Singh and Srivastava, 2021).

Brown and rainbow trout are evenly distributed in north western Himalaya, India. There distribution has expanded with human activities as well as by some environmental drivers (Singh and Srivastava, 2023) and the ecological and environmental changes determined with MaxEnt modeling has predicted habitat expansion of introduced non-native brown and rainbow trout in the Himalaya cautioning indiscriminate stocking of rainbow and brown trout in open waters such as river streams and lakes (Jan et al., 2023). With the habitat expansion of both the non-native trout in natural waters, there is a need of safeguard for local species and action plan for their sustainable management. Country's trout farming programmes need strengthening both in strength and scale against the current production of 2100 mt annually. The production could be raised upto 5000 mt by 2025. The success achieved in the sateṣ of Himachal Pradesh and Jammu & Kashmir, Uttarakhand and Sikkim in trout farming is worth emulating by other states for generating employment opportunities and adding trout fish in the food basket of people of hills. Due to the economic contributions of the growing aquaculture in coldwater sector, it is of great importance to consider the potential impact of climatic and demographic changes on the industry.

(ii) Chinese carp farming

Chinese carp farming in mid-hills is a popular practice among the farmers for the nutritional security as well for income generation. Due to simpler farming techniques, low input requirements and possibilities of integration of available resources, the culture of Chinese carp the grass carp *Ctenopharyngodon Idella* and silver carp *Hypophthalmichthys molitrix* and common carp *Cyprinus carpio* in small sized ponds (0.01, 0.03 ha) is becoming popular in the mid-altitudinal Himalayan region (Singh, 2015). The composition of cultured species varies from monoculture of common carp to polyculture of grass, silver and common carp. The carp culture in hills has been a proven more profitable culture system which can be integrated with dairy, horticulture, agriculture and paddy. Grass carp has emerged as a popular species for the low-cost hill aquaculture such as "Johra fishery" of north Bengal with higher realization of profit margins (Photograph 1). The developed package of practice of farming of Chinese carp in the mid hill region of IHR has increased the income of the marginal farmers.

Fig. 15.2: Commonly adopted Chinese carp and common carp farming practice in Himalaya

(iii) Improved strains of common carp

After long use of introduced common carp *Cyprinus carpio*, its performance in aquaculture deteriorated leading to stunting, reduced growth, early maturity and precocious breeding. Such observations of the stakeholders started drawing attention of the scientists and policy makers to deliberate upon further course of action particularly for introduction of improved strains of common carp. Consequently, improved Hungarian carp the "Ropsha scaly" and "Felsomogy mirror carp" were introduced for aquaculture in Himalayan region (Photograph 2 A & B). Improved Hungarian strains of common carp "Ropsha scaly" and "Felsomogy mirror carp" were introduced at Champawat Experimental Fish Farm of the ICAR-Directorate of Coldwater Fisheries Research (ICAR-DCFR) during 2007, cultivated and bred in hill region. After assessment and experimental studies, these two improved strains were found suitable for hill aquaculture due to their fast growth and tolerance to wide range of temperature (5-25°C). Males attained maturity in the 1st year, while females matured in 2nd year of rearing. There was 30-35% higher weight gain in "Ropsha scaly" and "Felsomogy mirror carp" during nursery rearing and 40-50% weight gain during grow-out phase with a survival of 80-82% over the performance of the existed common carp (Singh and Pandey, 2016). The improved strains "Ropsha scaly" and "Felsomogy mirror carp" have now been distributed to various Himalayan states such as Himachal Pradesh, Sikkim, Uttarakhand and J&K states of India for culture promotion.

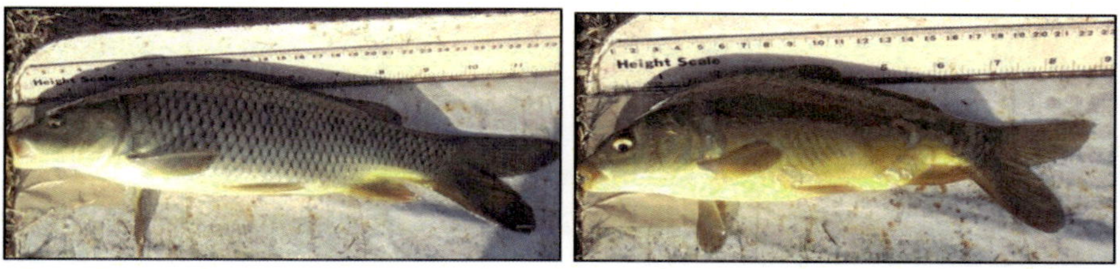

Fig. 15.3: A. Hungarian scale carp "Ropsha scaly" B. Hungarian mirror carp "Felsomogy carp"

(iv) Mahseer (*Tor putitora* and *Neolissochilus hexagonolepis*)

Golden mahseer (*Tor putitora*), the king of game fish is considered endangered fish. It is known for its excellent quality in terms of food value and behaviourally it is regarded as wonderful sports fish. The endangered golden mahseer, a Cyprinid fish species, is widely distributed across several south-east Asian nations (Sarma et al., 2016; Shahi et al., 2017; Pinder et al., 2019). Golden mahseer has been a vital source of food and recreation; however, the absence of standardized captive breeding protocol hampers the growth of the aquaculture potential of this fish species. Presently, seed production of golden mahseer is dependent on the annual gathering of gravid fish from the wild during the breeding season, which takes place between May and September. As a result, predictions regarding the quantity and quality of mahseer seed that will be produced for aquaculture are yet to be evaluated (Sarma et al., 2016; Akhtar et al., 2017).

Chocolate mahseer (*Neolissochilus hexagonolepis*) which is another important mahseer species has been reared successfully in captivity. Its captive breeding protocol has been standardized by the ICAR-Directorate of Coldwater Fisheries Research (ICAR-DCFR) (Mahapatra et al., 2011; Sarma et al., 2016; Dash et. al., 2021). The hatchery technology developed by the ICAR-DCFR has been now demonstrated in various states viz. Meghalaya, Nagaland, Mizoram and Assam for production of seed of chocolate mahseer for improving the status of aquaculture and fisheries (Sarma et al., 2016). It is believed that recent initiatives on culture and breeding of mahseer species both *T. putitota* and *N. hexagonolepis* will certainly help bring them in main stream of fisheries and aquaculture.

(v) Species diversification in coldwater aquaculture

Diversification of species is one of the strategies for sustainable development of aquaculture to boost the rural economy in upland region. Since, number of species cultured in coldwater sector is limited, there is a need to bring more number of fishes in aquaculture in order to increase the fish production in IHR. To meet the challenge, few promising species such as chocolate mahseer (*Neolissocheilus hexagonolepis*), and minor carp (*Labeo dero, L. dyocheilus, Osteobrama belangeri, Labeo pungusia* and *Bangana devdevi)* from different regions have been identified and assessed which could be cultured (Photographs 3 A & B) along with Chinese carp in mid-Himalayan region. It is expected that these species will play an important role to augment the coldwater fish production of the country.

Fig. 15.4:A. Labeo pungusia a potential candidate species of aquaculture

B. Banagana devdevi a potential candidate aquaculture species of north-east India

(vi) Sport fishing and eco-tourism

Angling or sports fishing is another most fascinating outdoor physical activities by the people of the Himalayan region. Now, it has become a form of eco-tourism promoting sustainable management of resource and consequently contributing to environmental conservation and recreational fisheries. Sport fishing is one of the leading adventure tourism activities across the world and in India for which the anglers travel to far distant places in order to satisfy their taste in angling anywhere and everywhere conferred with resources. Himalayan rivers and streams of India particularly of northwest, central and northeast Himalaya are home for native trout and mahseer fishes with excellent food and sports value. The people of the Himalayan region residing near the bank of rivers and streams would be much benefited if the mahseer fish and introduced non-native trout in connecting rivers are promoted. The sport fishing tourism is now developing in India, accounting for 2.7 % growth in tourism sector providing 31 million jobs in India. Today foreign tourist arrival is growing at an average of 4.5 % contributing foreign exchange earnings to the tune of over INR 1,35,193 crores in (Annual Report, Ministry of Tourism, 2015). There are three common methods for capturing mahseer: fly fishing, spoon fishing and bait fishing (Barua and Sarma, 2016). It is thus, important to take up promotional positive steps in this direction to adequately support sport fishing which could serve an alternate livelihood options for the people of the Himalayan sector.

(vii) Modern Aquaculture Tools

Modern aquaculture technologies such as adoption of Recirculatory Aquaculture System (RAS), aquaponics, biofloc system of rearing has prospects to revolutionize the coldwater aquaculture by allowing farmers to increase production, improve efficiency, and reduce environmental impact (Ying et al., 2015; Dash et al., 2018). Automated feeding systems, use of sensors and software to deliver precise amounts of feed to fish at regular intervals what we call as ‘precision aquaculture’ etc. These systems can reduce labour costs and prevent overfeeding, which can also overcome the problem of water pollution and fish health issues. Selective breeding programs are further used to improve the performance and traits of fish species used in coldwater aquaculture. Such programs can lead to faster growth rates, disease resistance, and improved flesh quality.

Challenges

At present attention is being given towards the standardization of technologies for various coldwater fish species suitable to the Himalayan region depending on investment potentials. This also focuses on species-based systems, as a high biodiversity of fish species exist in the Himalayan region of the country (Singh and Akhtar, 2015). With the growing scientific practices coldwater aquaculture has flourished and now achieving new heights of production; however, farmers are facing new challenges of diseases. Concurrently, adoption of fish health management approaches need to be advanced to reap healthy and quality aquaculture production. The biotechnological approach has shown promises both to improve the health status of cultured fish and to safeguarding the aquatic environment (Bhat et al., 2023). Coldwater aquaculture relies heavily on sex as an essential trait, since several species display relatively faster growth in either female or male requiring discoveries of critical miRNAs during sex differentiation in the testes and ovaries. With the advancements in sequencing technology and bioinformatics, it will be exciting to investigate the function and importance of different types of molecules, such as circular RNAs (circRNAs), which govern numerous biological functions. The following points need be deliberated in order to improve breeding and culture of fish in the hills:

- Imparting training to farmers and entrepreneurs of the mountain at different levels along with better co-ordination between extension functionaries.
- Development of suitable and specific aquaculture system for micro climatic conditions of the mid and high altitude areas.
- Regular monitoring of reproductive biology of economically important fish species of Himalayan region so as to suggest remedial measures for depleting population.
- Conservation and rational exploitation of resources, which are the backbone of fish and fisheries in the Himalayan region particularly at the high altitude areas.
- Introduction of research programmes aiming at intensification and diversification of freshwater aquaculture in tune with the geomorphological feature of the region.
- Extensive conservation and bio monitoring of all coldwater resources in time and space to assess threat perspectives in relation to biodiversity.

- It is important to explore new candidate fish species for coldwater aquaculture.

Coldwater fisheries and aquaculture are threatened by the destruction of habitats such as rivers, lakes, and wetlands. Dams, water withdrawals, pollution, and other human activities can disrupt or destroy these habitats, reducing the availability of suitable habitat for coldwater fish (Sarma et al., 2018). Overfishing can deplete populations of coldwater fish and disrupt the balance of entire ecosystems. In addition, overfishing can lead to the collapse of commercial and recreational fishing industries. Non-native species that are introduced into coldwater habitats can outcompete or prey on native fish, altering the entire ecosystem. Invasive species can adversely impact the native species diversity and ecosystem services (Singh, 2021). Poor water quality, including high levels of pollutants, can harm coldwater fish and their habitats. Chemicals, nutrients, and other pollutants can affect the health of fish and other aquatic organisms, and can lead to the proliferation of harmful algal blooms. The public may have concerns about the environmental impacts and sustainability of aquaculture, particularly in areas where it is a new and emerging industry. Overall, the prospects for coldwater aquaculture are positive, but addressing the environmental constraints will require ongoing research, innovation, and a commitment to manage the sustainable practices.

Future projection and way forward

So far governance for aquaculture of potential Coldwater fish species such as trout, carp and mahseer is very limited and state governments do not have any mandatory mechanism to extend technical backstopping to the private aquaculturists. Nevertheless, some necessary governance issues are flagged below to cater the immediate need:

i. The regulations governing site selection and construction of trout farm particularly has not been robust and environment friendly, and hence it may disturb overall maintained functionality of the habitats where trout farms are sited. It is therefore, the government departments and research institutes should take up responsibility to identify suitable areas for trout culture in IHR.

ii. Some farming standards is required to be developed for implementation with regards to water recirculation, stocking density and size, feed quality, management practices, disease management and brood raising. A detailed developed guideline should be made available to the farmers and stakeholders.

iii. Since the intensification of trout farming is on rise, it should be ascertained that chemicals and pollutants if present in trout culture used water (waste water) should not disturb ecosystem functionality and in turn affect habitats. Although there is hardly any such deviation studied or reported from India, yet it is important that the research institutions should develop a standard on water quality especially on the water chemistry of used water from trout farms so as to develop a model for implementation.

iv. A robust and environment friendly regulatory governance for trout farming is needed that will overall maintain functionality of the habitats where trout farms are sited. Potential success in trout production could be achieved through improved governance and significant improvement in the management practices.

Conclusion

Coldwater aquaculture is an important industry, and the most commonly farmed coldwater species in India include rainbow trout, carp, and mahseer. These species are typically grown in freshwater ponds, tanks, or cages in coldwater environments. The total volume of coldwater aquaculture production in India is small. However, the use of advanced technologies, such as RAS and closed containment systems of aquaculture; fish health management approaches; brood stock management and biotechnology can be significantly used to increasing overall production. These technologies allow for better productions, control of water quality and temperature, and can reduce the risk of disease and pollution. Many Asian countries are focused on exporting their coldwater aquaculture products to international markets, particularly Europe and North America. The sustainability of coldwater aquaculture in India is a concern, particularly with regard to the impact of fish farming of non-native fish species on the aquatic environment and wild native fish populations over which thousands of marginal fishermen depend for their livelihood and subsistence. Efforts are being made to improve sustainability, such as the development of eco-labelling programs and certification schemes in the interest of the farmers and entrepreneurs.

There is an increasing recognition that trout aquaculture can contribute to blue and green growth and the bio-economy of mountain people. The use of sustainable resources, food security and public health by providing sustainably produced trout of high quality as tagged health food and other native fish species is the need of the day. Identifying challenges for sustainable development of non-native trout culture, and hence growth, has to refer to an extensive combination of economic, social and technical issues (Singh, 2020; Singh, 2021). Whilst many scientific approaches apply specifically to the coldwater aquaculture development, many are more general in nature, covering rearing conditions, water use, environmental considerations and supply chain in general while others require modern aquaculture tools. The complex interweaving of governance and societal approaches that cover these issues tends to show that, until recently, aquaculture of trout has been an 'add-on' that remained outside core policy considerations in Himalayan region.

Challenges for sustainable development of trout, carp and mahseer culture need to be conducted conveying responsibility of following good governance and technological standards. Standardized disease management mechanism in trout culture their governance on stocking of non-native trout and carp in open waters must be strengthened and shared with stakeholders. Certain state waters should be designated as heritage trout and mahseer waters to exclusively manage wild species and domesticated fishes which should provide anglers an opportunity to fish in aesthetically pleasing and environmentally productive waters with appropriate angling regulations. Himalayan Water is effectively the habitat where rainbow trout and mahseer live so it must be monitored and preserved as best as possible in terms of physico-chemical parameters and water quality. All these aspects are the basis of ensuring high-quality trout farming and mahseer rearing. Sustainable management, together with a genetic program of rainbow trout and mahseer should be employed, which will be the next challenge to further improve the fish performance under changing environment in the time of climate change. Institutional approaches to offer a comprehensive frame work on good governance and technological standards should be ensured. There should be a scientific stocking policy for production augmentation and also enhancing the capacity to mitigate climate change impacts which is a critical challenge. For all issues, there should be a collaborative effort among governments, research institutions, industry, stakeholders, to ensure the effective support and development of coldwater fisheries in the Himalayan region.

References

Akhtar M S., Ciji A., Sarma D., Rajesh M., Kamalam B S., Sharma P., Singh A K. (2017). Reproductive dysfunction in females of endangered golden mahseer (Tor putitora) in captivity. Animal Reproduction Science Appeared online 18 May, 2017 https://doi.org/10.1016/j.anireprosci.2017.05.004.

Barua D and Sarma, D. 2016. Mahseer in recreational fisheries and ecotourism in India. Published by ICAR-Directorate of Coldwater Fisheries Research, Bhimtal (Nainital), Uttarakhand. India Bulletin No. 26, pp. 40.

Bhat Raja Adil, Mallik S K., Tandel R S., and Shahi N. 2023. An Overview of Coldwater fish diseases and their control measures. In: Fisheries and Aquaculture of the Temperate Himalayas, P. K. Pandey et al. (eds.). 255-283. https://doi.org/10.1007/978-981-19-8303-0_15.

Dash P, Pandey N N., Mallik S.K., Ritesh Tandel, Singh A K and Raja Adil (2018). The addition of probiotic bacteria to microbial floc: water quality, growth, non-specific immune response and disease resistance of Cyprinus carpio in mid-Himalayan altitude. Aquaculture 495, 961-969.

Dash P., Tandel R S., Bhat R A H., Sarma D., Pandey N N., Sawant P B., Chadha N K. 2021. Spawning substrate preference and spawning behavior of chocolate mahseer, Neolissochilus hexagonolepis. Animal Reproduction Science 233, 106847, ISSN 0378-4320, https://doi.org/10.1016/j.anireprosci.2021.106847.

Jan, A., Arismendi I., Giannico G., and Flitcroft R. 2023. Habitat overlap among native and introduced coldwater fishes in the Himalayas. Scientific Reports 13:15033. https://doi.org/10.1038/s41598-023-41778-y

Mahapatra B K and Vinod. K. 2011. Reproductive biology and artificial propagation of chocolate mahseer Neolissocheilus hexagonolepis (Mc Clelland) in Meghalaya, India." Indian Journal of Fisheries 58: 35-40.

Pandey N N and Singh A K. 2018. Breeding and culture of rainbow trout, Oncorhynchus mykiss. In: Package of practices for breeding and culture of freshwater fish species published by National Fisheries Development Board, Hyderabad p37-44.

Pinder A C, Britton J R, Harrison A J, Nautiyal P, Bower S D, Cooke S J, Lockett S, Everard M, Katwate U, Ranjeet K, et al. 2019. Mahseer (Tor spp.) fishes of the world: Status, challenges and opportunities for conservation. Rev. Fish Biol. Fish. 29: 417–452.

Pramanik P, and Bhaduri D. 2016. Impact of Climate Change on Water Resources in Indian Himalaya. In: Bisht, J., Meena, V., Mishra, P., Pattanayak, A. (eds) Conservation Agriculture. Springer, Singapore. https://doi.org/10.1007/978-981-10-2558-7_19.

Sarma D, Singh A.K. and Akhtar M.S. 2016. Mahseer in India: Resources, Breeding, Propagation, Conservation, Policies and Issues. Published by ICAR-Directorate of Cold Water Fisheries Research, Bhimtal, 168pp. ISBN: 978-81-931230-0-3.

Sarma D and Singh A K. 2018. Hatchery management and seed production of golden mahseer, Tor putitora. In: Package of practices for breeding and culture of freshwater fish species published by National Fisheries Development Board, Hyderabad p29-35.

Shahi N, Singh A K, Sahoo M, Mallik S K, Thakuria D. 2017. Molecular cloning, characterization and expression profile of kisspeptin1 and kisspeptin1 receptor at brain-pituitary-gonad (BPG) axis of golden mahseer, Tor putitora (Hamilton, 1822) during gonadal development. Comp. Biochem. Physiol. Part B Biochem. Mol. Biol. 205: 13–29.

Singh A K, and Lakra, W S. (2011). Risk and benefit assessment of alien fish species of the aquaculture and trade into India. Reviews in Aquaculture, 3: 3-18.

Singh A K, and Pandey N N. 2016. Aquaculture prospects of introduced improved Hungarian strain of common carp in Himalayan region, India. Published by ICAR-DCFR Bulletin, 18 pp.

Singh A K. 2015. Advances in Indian Coldwater fisheries and aquaculture. J Fisheries Sciences.com 9(3): 048-054 (2015).

Singh A K. 2020. Emerging scope, technological up-scaling, challenges and governance of rainbow trout Oncorhynchus mykiss (Walbaum, 1792) production in Himalayan region. Aquaculture 518, 734826 https://doi.org/10.1016/j.aquaculture.2019.734826Wang, 1998.

Singh A K. 2021. State of aquatic invasive species in tropical India: an overview Aquatic Ecosystem Health & Management, 24(2): 7–17, 2021. ISSN: 1463-4988 print / 1539-4077 online. DOI: 10.14321/aehm.024.02.04 .

Singh A K. and Akhtar M S. 2015. Coldwater fish diversity of India and its sustainable development. Biodiversity for Sustainable development Editor-in-chief Pratibha Singh Published by UP Biodiversity Board, Lucknow. International Day for Biological Diversity on 22 May, 2015. pp 97-105.

Singh A K. and Srivastava S C. 2021. Improved feeding strategy to optimize growth and biomass for up-scaling rainbow trout Oncorhynchus mykiss (Walbaum 1792) farming in Himalayan region. Aquaculture 542, 736851http://doi.org/10.1016/j.aquaculture.2021.736851.

Singh A K and Sarma D. (Eds) 2017. Aquatic resources and fish diversity of the Himalayas. Narendra Publishing House, New Delhi 345pp. ISBN 978-93-86110-81-7.

Singh A K and Srivastava S C. 2023. Environmental drivers inducing habitat expansion and shift of introduced alien trout in the Himalayan ecosystem and management concerns. Journal of Nature Conservation 73, 126393. Doi.org/10.1016/j.jnc.2023.126392.

Ying L, Baoliang L, Ce S, Guoxiang S. 2015. Recirculating Aquaculture Systems in China-Current Application and Prospects. Fish Aquac J 6:134. DOI: 10.4172/2150-3508.1000134.

Zaidi S G S, Singh A K and Sarma D. 2018. Breeding, seed production and rearing of coldwater ornamental fishes in aquarium. ICAR-DCFR Bulletin No. 27. Published by Directorate of Coldwater Fisheries Research, Bhimtal, Nainital, Uttarakhand, India pp-33.

16

Entrepreneurship Opportunities in Fish Feed and Food Sector in India

K. Ambasankar and V. Rajaram*

Central Institute of Brackishwater Aquaculture 75, Santhome High Road R.A. Puram, Chennai 600 028
**Grobest Feeds Corporation, 25, Grobest Towers, 1A1B, Perambur Red Hills High Road, Vinayakapuram, Chennai,-600099*

Abstract

Fish feed is one of the most important factors for successful and lucrative fisheries. This chapter provides a overview of the fish feed sector, the demands and also discusses the entrepreneurship opportunities in fish feed sector.

Keywords: *Fish Feed, Feed in aquaculture, Aqua feed mill, sinking feeds, floating feeds, Alternative plant proteins*

Introduction

Aquaculture and Fisheries remain important sources of food, nutrition, income, and livelihoods for hundreds of millions of people around the world including in India. Total fisheries and aquaculture production reached a record of 214 million tonnes in 2020, comprising 178 million tonnes of aquatic animals and 36 million tonnes of algae (FAO 2022). Global aquaculture production continued to grow amid the worldwide spread of the COVID-19 pandemic. Global aquaculture production consisted of; 87.5 million tonnes of aquatic animals predominantly used as food for human consumption; 35.1 million tonnes of seaweeds and other micro-algae for both food and non-food uses. Global consumption of aquatic foods has increased significantly. In 2019, global aquatic food consumption was estimated at 158 million tonnes and consumption increased at an average annual rate of 3.0 percent from 1961, compared with a population growth rate of 1.6 percent. In India, the total fish production during the year 20-21 is estimated at 14.73 Million Metric Tonnes (MMT) with a contribution of 11.25 MMT from the inland sector and 3.48 MMT from the marine sector. The fisheries sector contributes about 1.24% to the gross value added (GVA) and 7.28% to the agricultural GVA. In India, the changing consumer preferences, advancements in technology and income growth are the real drivers that augur well for further expansion and demand of food from aquatic origin. Against this backdrop, this sector is poised for unprecedented growth and opening various avenues of entrepreneurship.

Aquaculture

Freshwater carps and brackishwater shrimp are the dominant cultured candidate species in India. Freshwater aquaculture contributes about 90% of the total aquaculture production. The farming of Indian Major Carps (IMC) is the mainstay in fresh water. Though the production is predominantly traditional the conversion to feed-based aquaculture is getting momentum. However, feed-based aquaculture is growing rapidly for the species like tilapia, pangasius, and pacu. Shrimp farming is the face of brackishwater aquaculture in India and shrimp culture has been carried out as a traditional activity for ages in India. For a developing country like ours, shrimp farming is a high-potential sector with enormous scope for increasing foreign exchange and employment generation. The early nineties witnessed a phenomenal growth of the sector which was entirely dependent on the culture of a single species, the tiger shrimp, *Penaeus monodon*. During this period, the shrimp culture was a low-risk, high-profit venture. In the late 90s, there were serious problems with viral diseases and environmental safety issues, which arose mainly because of the lack of planning and regulation. The Central Institute of Brackishwater Aquaculture (CIBA) and National Bureau of Fish Genetics and Resources (NBFGR) aimed at evaluating the feasibility of the introduction of this new species and carried out the risk analysis for introducing the exotic shrimp, *Penaeus vannamei*. Since the introduction of vannamei in 2009, Indian shrimp farming has witnessed a rapid shift towards the culture of vannamei almost replacing tiger shrimp, except for the traditional culture practiced in West Bengal and Kerala, where still monodon and indicus are farmed. To sustain the rapid growth of vannamei farming, the feed industry also needs to support the sector with quality and cost-effective feeds, as feed alone contributes 50-60% of the operative cost in the semi-intensive shrimp culture practiced adopted by Indian farmers.

Significance of feed in aquaculture

The significant increase in aquaculture has been possible with the corresponding increase in the availability of formulated feeds. Aqua feeds have been developed for sustainable and commercial culture fish and shrimps to meet the animal protein needs in developing countries. The principal cost in the manufacture of aqua feeds is that of raw materials; this could amount to as much as 80 percent, or more, of the manufacturing costs in commercial feed mills. Aqua feeds in India have evolved commercially in the late 90s for shrimp culture and then for fish during the last decade. In the brackishwater aquaculture sector majority of the feed used is scientifically formulated compounded feed produced by multinational / Indian companies. There are some small players with a capacity to produce 1 tonne per hour are entering the market. In freshwater aquaculture, the traditional farm-made feed consisting rice bran with any of the oil cakes available in their locality. The sinking feed for carps and the availability of floating feeds for fish seems to make a steady and positive impact in changing the traditional feeding system and increasing productivity. However, farmers are more concerned about the cost of production and the final product quality and the majority of the farmers are using the formulated feed as a supplementary feed and not as complete feed. Here it is pertinent to note that the price of DORB has a direct bearing on the use of formulated feed. Currently, in freshwater aquaculture, extruder feeds are popular for species like pangasisus, tilapia, and pacu. Though the traditional bag feeding technique is followed for IMCs, about 15% of the production is under sinking and floating extruded feeds and thus opening big opportunities for the formulated feed market to grow in the freshwater sector also increasing aquaculture production through the feed.

Most of the corporate and big feed mills are having the state of the art facilities for the production of quality water stable shrimp feed pellets. The shrimp feed mills have facilities for coarse grinding, fine grinding, sieving, mixing and steam pelleting with three-stage pre conditioner combined with post-pellet conditioner is in vogue. Most of the shrimp feed produced in India uses a ring die pellet mill to produce a sinking compact pellet. In general, the shrimp feeds for tiger shrimp consist of three grades of crumble (C1, C2 and C3) and three to four grades of pellets in the diameter of 1.8, 2.0 and 2.2 mm. However, there is a gradual shift in sizes of the pellet in the feed meant for pacific white shrimp, *Penaeus vannamei*. Now the feed pellet sizes available as 1.0, 1.2, 1.4, 1.6, 1.8 and 2.0 mm are being used for feeding vannamei (Ambasankar et al., 2017). Though the production of feed pellets at sizes less than 1.8mm is a high energy consumption process, feed millers and farmers are of the view that smaller pellets are better and could able to meet the requirement of the current practice of white shrimp farming at higher densities than *P.monodon*. The inherent advantage of smaller diameter pellets is that crumbling is a two-step process wherein bigger pellets (1.8 mm) are subjected to crumbling to get the required size (400 microns to 1.7 mm) and the time delay in crumbling operation is reduced in case of taking production with smaller size dies. Attempts have been made to produce extruded shrimp feeds with reasonable success and few firms started producing the extruded shrimp feeds.

Aqua feed mill as opportunities for entrepreneurship

Since the introduction of vannamei, Shrimp production in India is growing every year irrespective of some diseases, climatic conditions, and price volatilities. Feed demand from shrimp and prawn farming is more than 1.4 million metric tons a year, while a near similar tonnage of fish feeds is also produced in India from the 40 plus shrimp and fish feed mills with a production capacity of >3.0 million tons. Still few companies are expanding their capacities and new companies are coming up. Most of the feed mills are situated in Andhra Pradesh and Tamil Nadu. There is a huge scope for establishing small to medium-scale feed mills in the inland states for freshwater aquaculture feeds and also opportunities for the other coastal states to establish a feed mill to cater to the need of locals. The marketing channel of feed distribution from the feed mill to the dealer, sub dealer/distributor and then to the farmer along with increased freight costs makes the feed cost higher for farmers in many states. At this juncture a small to the medium-scale feed mill to cater to the need of small and medium farmers with the concept of factory to farm will help in many ways. ICAR- CIBA already established such small feed mills and they are running successfully. These types of feed mills have to be located in each aqua farming zone so that the farmers could directly get it from the feed mill for an affordable cost.

The increased demand for feed processing is being contributed by shrimp feed equipment manufacturers from international as well as Indian feed mill machinery manufacturers. Big feed mills can be established by providing turn-key projects and arranging technical consultancies while small feed mills can be established as per the customer requirement taking into consideration of the

Establishing a feed mill

In aqua feeds there are two types of feed mills. The feeds for fish are produced through an extruder (Figure 16.1) and pelletizer (Figure 16.2). If the requirement is for floating fish feed, (Figure 16.3) extruder is the option, and if sinking feed (Figure 16.4) is also required an option of pelletizer has to be additionally incorporated. Crustacean including shrimp and crab requires a sinking feed and the ring die pelletizer is normally used to prepare the sinking feeds. If both feeds are to be produced any integrated feed mill encompassing an extruder and pelletizer can be established..

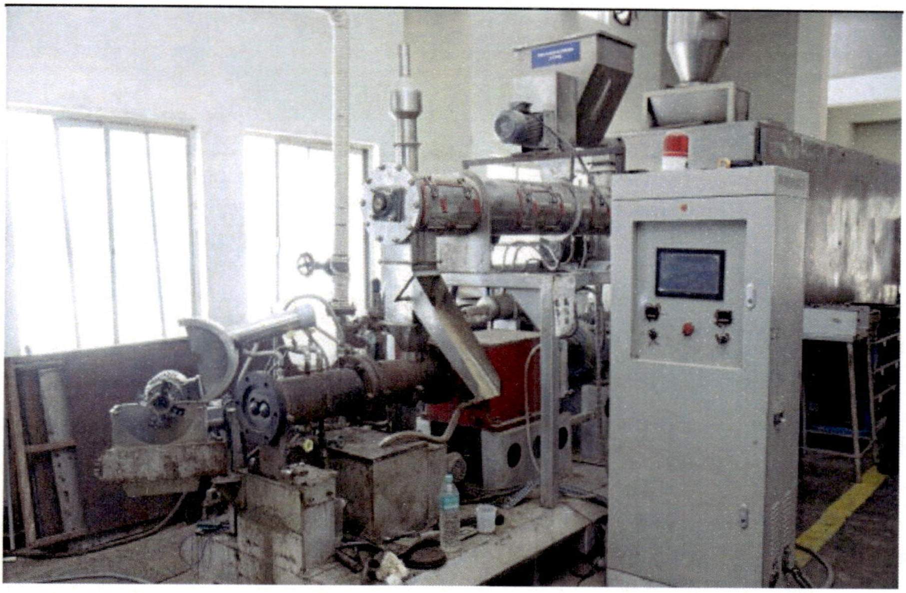

Fig 16. 1: Pilot scale twin screw extruder for floating shrimp feed production

Fig. 16.2: Pilot scale ring die pelletizer for fish feed production

Fig. 16.3: Floating feeds for fish produced using twin screw extruder

Fig. 16.4: Sinking feeds for shrimp produced using ring die pelletizer

Feed mill Site selection

The site should be preferably away from the residential area and should be well connected with the road. The location should be free from probable social issues, clear documents, government support, availability of required power, etc., ideally industrial development land allotted by state governments are more suitable compared to other options. The statutory requirement to establish any industry has to be obtained for feed mills also.

Cost of the project

The cost of the project will depend on the level of production and the level of automation expected. ICAR- CIBA has got the expertise in establishing small and medium types of feed mills with an investment capacity of ₹ 60.0 Lakhs to ₹ 6.0 Crores. Normally the cost of the machinery for the production of 1 ton per hour of shrimp/fish feed will be to the tune of 1.5 crores.

Raw materials used in aqua feed production

India is rich in agricultural products and its by-products and the array of raw materials available can cater to the demand of the industry. Most of the raw materials are available locally for shrimp and fish feeds. However, the prices are highly volatile and sometimes higher than international prices. Hence, some of the critical raw materials like fish meal, krill meal, additives, minerals and vitamins are being imported. The major food commodities produced in India are given below in Table 1 (FAOSTAT 2019).

Table 16.1: Major food commodities production (2019).

S.No	Commodity	Million MT
1	Rice	177.6
2	Wheat	103.6
3	Maize	27.7
4	Cotton Seed	18.6
5	Soybean	13.3
6	Coconuts	14.7
7	Millets	10.2
8	Chick Peas	9.9

Fishmeal

India produces about 2.5 to 3.0 lakh metric tonnes of fish meal and 20 - 30 thousand metric tonnes of fish oil, every year which is sufficient for the feed industry, but the availability and catch of fish are highly fluctuating, so companies import fish meal from Peru, Chile, Canada, USA, Oman, Mauritius and Africa

Soybean meal

Recorded soybean seed production in India is 12 million MT. 1 million MT are kept for sowing and about 8 million MT of soybean meal is produced, of which

5 million MT goes to the poultry industry, 0.7 million MT to aquaculture and 2.3 million MT is left for human consumption and export..

Alternative plant proteins

Rapeseed meal, groundnut/peanut meal, cotton seed meal, corn gluten, Distillers Dried Grains with Soluble (DDGS), sunflower meal, wheat gluten, rice Gluten, rendered terrestrial animal proteins like blood meal, meat and bone meal, hydrolyzed feather meal, poultry by-product meal are also available for aqua feeds. The list of raw materials used in shrimp and fish feeds in India is given below in Table 16.2.

Table 16.2: Commonly used raw material cost and their sources in fish and shrimp feeds

Raw materials	Source*	Price (INR/KG)	
		Low	High
Fish meal (import)	CL & PR	105	140
Fish meal 67	KA, KL & TN	90	120
Fish meal 60	KA, KL & TN	88	100
Squid meal	GJ	90	120
Squid liver powder (import)	SK	80	90
Squid liver powder	KL, SK	65	70
Acetes sp	MH	45	65
Shrimp shell meal	TN, KL, AP	42	55
Blood meal	UP,MP & TG	45	60
Soya Ext. 48	MP & MH	38	50
Soya Ext. 50 (Hi pro)	MP & MH	40	65
Soya flour	MP & MH	40	65
Wheat gluten meal	PB & NOR	180	265
Corn gluten meal	MH	65	75
Rape seed meal	RJ, MP & WB	25	40
Ground nut meal	KA,TN & AP	35	55
Cotton seed cake	MH, GJ & MP	30	50
Dried yeast	BR	70	90
Maida	PB, AP, TN, MH & KA	23	32
Broken rice	AP & TN	19	26
Maize	AP, KA & TN	14	26
Bajra/Pearl millet	RJ & UP	14	26
Raw rice bran	AP, MH, UP & WB	20	35
De-oiled Rice bran	AP, MH, UP & WB	14	22
Fish oil	KA, KL & TN	140	240
Lecithin	MP, MH	90	200
Squid oil	SK	200	300

*AP-Andhra Pradesh; GJ-Gujarat; MP-Madhya Pradesh; KA-Karnataka; MH-Maharashtra; RJ-

Rajasthan; KL-Kerala; PB-Punjab; SK-South Korea; UP-Uttar Pradesh; NL-Netherlands; BR-Brazil; NOR-Norway; TG- Telangana, CL-Chile; PR-Peru

Energy sources and its by-products

Cereal grains like wheat, rice and maize are the major energy grains used in the aqua feed and millet can also be used as a source of energy. Rice bran and de-oiled rice bran and Rice polish are the important sources of energy used in aqua feed and these energy by-products from the rice and wheat milling industry are available for use in the aqua feed industry

Lipids

Most aquaculture species require the essential fatty acids eicosapentaenoic acid (EPA) and docosahexaenoic acid (DHA) and phospholipids for normal growth, vitality, immune and other physiological functions. Fish oil with high content of essential fatty acids (EPA & DHA) and soybean lecithin with phospholipids are used as primary lipid sources. Now algae oil and other vegetable oils are used as lipid sources.

Feed manufacturing Methods

Aquaculture feeds are generally manufactured by pressing the finely powdered ingredients with a ring die or plate dies to form a cylindrical shape pellet, usually, 1-2 times the diameter for floating feeds and 2-4 times the diameter for sinking feeds. Conventional pelleting provides only sinking pellets for shrimp and crabs. Whereas extruders emerge strongly due to their versatility in using a wide range of raw materials and producing both sinking and floating pellets.

i) Conventional Pelleting (sinking feeds)

(i) Extruder (sinking, slow sinking and floating feeds)

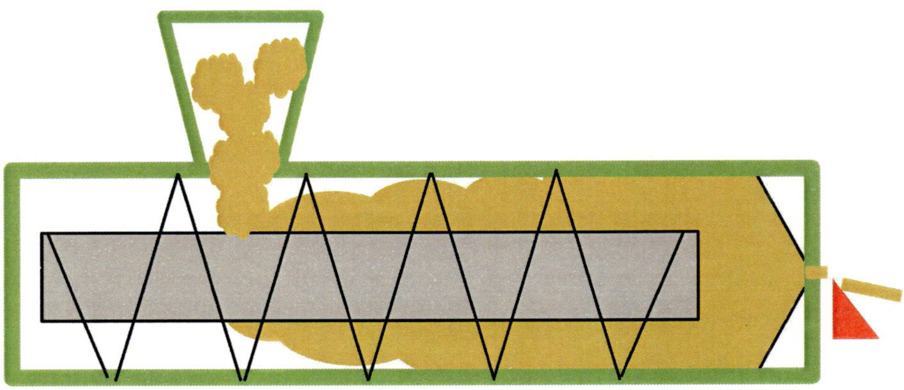

Fig. 16.4: Relevance of micro feeds for shrimp larvae in the Indian context

The majority of the hatchery feeds, mainly larval feeds for shrimp and fish in the Indian market are imported ones. To our knowledge, currently there is no known larval feed manufacturing firm that produces larval feeds on a commercial scale. At present the larval feeds used in hatcheries are imported and from multinationals. The cost of larval feeds ranges from INR 2000 to 3000 per kilogram. As per the industry estimate about 700 to 1000 tons of larval feeds are used per annum and the feed cost alone comes to about 300-400 crores per annum. There is a bright scope for entrepreneurship in larval feed production and marketing. ICAR- CIBA has done extensive research in larval feed and has developed the larval feed for shrimp, CIBA Shrimp larvi^{Plus} This feed has been tested in the commercial vannamei hatcheries in collaboration with a private partner and the results showed on par performance with imported feeds. The main advantage of the Shrimp larvi^{Plus} is its cost-effectiveness and this feed is in the process of commercialization. There is good scope for indigenous larval feed manufacturing in India and CIBA can give technical support in this technology.

Nursery Feeds

Farmers are now exploring innovative approaches to shrimp farming. One of the latest developments "on-farm nursery," where shrimp post-larvae are raised in a bio-secured and controlled environment similar to larval rearing in a hatchery. High quality feeds for the nursery are essential for its success. Currently, we are producing 120 billion shrimp larvae per annum and even if 10% of PLs are grown in the nursery, it will require 3000 tonnes of feeds. The prices of nursery feed range from INR400 to 1000 per kg and this opens up a new feed segment called "Nursery Feeds" with huge growth potential.

Micro feeds production

Micro feeds are characterized by their sizes ranging from 50 microns to 900 microns and prepared in a form that is most suitable for the larvae of fish and shrimp to consume and could able to retain the nutrients with suitable buoyancy for the species concerned. These are prepared as sinking, floating and slow sinking feeds of required particle size (50- 900 microns) and are produced for a highly specialized market. These are presently produced as

1. Micro-particulated feeds - Production of large pellets (1 to 1.2mm) and crumbling
2. Direct extrusion of small diameter feeds of 500 to 900micron
3. Micro extruded and marumerized
4. Spray-dried (50- 200 microns)
5. Micro coated and Micro encapsulated

Microparticulated dry diets

These diets are prepared in dry powder form. The particle size of the diets varies from 50 to 500 microns. A variety of ingredients with different functional and nutrient properties are used for formulating larval diets. The selected quality ingredients are arranged in the form of a balanced feed formula. After finely grinding the individual ingredients they are mixed according to the formula. A suitable binder is also included for making the diet stable in water. The diet is prepared as pellets or cubes or flakes and dried at low temperatures as for as possible. The dry diet is then reground and sieved to obtain the desired particle size. These micro particulated dry diets are used for feeding larvae. Separate microparticulated diets were developed for fish and shrimp larvae. In general, larval diets are formulated to contain more natural raw materials such as premium fish meal, fish protein hydrolysate, fish oil, prawn head powder, cereal flours and feed additives. Starch present in cereal flours is mainly intended as a natural binder by the gelatinization during cooking and good water stability of the feed.

Direct extrusion of small diameter feeds of 500 to 900 microns

Micro feeds in the sizes of 500 to 900 microns are directly extruder in a specialized micro extruder and the feeds of floating, slow sinking and sinking can be produced. These require special machinery for grinding, mixing, preconditioning, cooking, extrusion and post-pelletizing. Moreover, all the processing parameters such as moisture, starch level, the fat content of the formula, ingredients combinations, the temperature of cooking, pressure in the extruder barrel, etc. needs to be optimized for each formula.

Micro extruded marumerized (MeM) diets

The production of MeM diets involves the technology of cold micro-extrusion and marumerization (Figure 5) and can ensure high water stability of soluble and insoluble nutrients in the feed while avoiding the use of chemical binders. A uniformly mixed and fine ground formulation is passed through a low shear and low-temperature extrusion process where it is conditioned with steam, water and other possible liquid additives and compressed through a special dome die to form micro extruded strands. These strands, when delivered into a spheronizer spins in a cyclonic motion, the strands will be broken into small pieces and made into spheroids in a few seconds. The sizes and shapes of the pellets will be determined by the initial diameter of the extruded strand and the configuration of the spinning disc in the Marumurizer. Feeds in the size range of 0.3 to 1.2 mm can be produced. Low processing temperature minimizes nutrient damage in this way of processing. Here the advantage is cold micro-extrusion avoids high temperature during the production process and keeps high vitamin levels and digestibility of nutrients and favours the production of medicated feeds and utilization of other temperature sensitive ingredients.

Fig. 16.5: Cold extrusion cum Spheronization (Micro extruded marumerized feed production)

Spray dried diets

The finely powdered feed mix is mixed with the required quantity of water and then spray dried using a spray drier. The spray dried particle sizes are controlled by varying the atomization, flow rate, and temperature. Microencapsulation is also possible through spray drying by coating the spray dried powder with proper binders. These are specially used to prepare feed particles in the size range of 50 to 200 microns. The main disadvantage of this method is it is not amenable to all the materials and certain heat sensitive nutrients are lost during this process.

Micro coated and microencapsulated diets

Micro coated diets are obtained by coating the micro particulated diet with suitable materials. This is done mainly to prevent the leaching of nutrients from diet particles and to improve the nutritive value of the diet. Cholesterol, lecithin, zein and nylon protein are used as coating materials.

Shrimp larval diets are prepared as capsules to prevent the leaching of important nutrients into the water before entering the stomach of larvae. Since these capsules are minute in size (10 to 200 microns), they are known as microcapsules. Different types of diets such as liquids, semi solid and dry powders can be encapsulated. The material used for making the capsule is called the wall and the inner portion (diet) is called the core material. The principle involved in encapsulation is interfacial polymerization. Many materials are used as wall coatings. These are gelatin gum acacia, egg albumin, glucopeptides, chitosan and nylon protein

Organic feed

Organic farming delivers the highest quality food, produced without artificial chemicals, pesticides or genetic modification, and with due consideration for animal welfare and the environment. At the same time, it preserves the landscape and non-interferences with the rural communities. Organic food production promotes biodiversity, biological cycles and biological activity. It aims to manage food production as an integrated, whole system and encourages the maintenance and sustainability of this system by restricting the introduction of harmful substances. The usage of organic feed is the prerequisite for organic aquaculture production and this niche area is also expected to grow in the coming areas and would pave the way for entrepreneurship

Organic farming is another way of adding value to the produce. This is also another form of certification scheme where the standards specify the use of organically produced inputs and maintenance of the welfare of the animals with minimal disturbance to the ecological conditions of the farming area. Though organic agricultural products are already being produced and marketed, it is still in a very formative stage in aquaculture. CIBA in Collaboration with M/S Jass venture private limited demonstrated the culture of organic shrimp and the product has been marketed as organic shrimp. Presently Agriculture Products Export Development Authority (APEDA) is developing guidelines for certifying organic shrimp and once these are accepted by other countries, then the cost of organic certification would be reduced.

Functional feeds for health and growth

Diseases are the major bottleneck in aquaculture and major emphasis is given to prevention rather than the therapeutic approach. In this direction, functional feed targeting improved health and disease residence and improved growth and coloration are the important endpoints the industry is looking for. In this direction the functional feed market

A functional feed is defined as a feed that provides superior performance than that achieved using a regular feed by incorporating specialty ingredients that promote growth, health, and survival (Soto et al. 2015). Diseases are the major bottleneck in aquaculture and major emphasis is given to prevention rather than as therapeutics. Hence, the development of functional performance feeds represents a huge opportunity to increase the productivity, profitability, and sustainability of the aquaculture sector. A wide range of ingredients are used to prepare functional foods such as nutraceuticals, pigments, vitamins, minerals, omega-3 rich oils, plant extracts, fermented products, essential oils, enzymes, probiotics, prebiotics, etc, are specifically targeting to improve growth, colouration, immune function, digestion, meat texture, omega 3 fatty acids and overall well-being beyond meeting basic nutrition needs. Functional feed is perfect for someone who wants to create a unique and innovative market segment that offers excellent growth potential in addition to being sustainable.

Feeds for aquariculture

Aquarium keeping hobby is on the rise across the world, and the ornamental aquarium industry is today a multi-billion dollar industry with an estimated value of 15 billion US dollars involving trade in over 5000 species of fish. An ornamental fishery is the second most popular hobby with about 100 million hobbyists in the world. India is blessed with a rich natural biodiversity of fish species including over 400 ornamental fish species. In addition to international trade, experts estimated that the Indian domestic aquarium market will grow from Rs. 300 crores to Rs. 1200 crores. Feed alone contributes about 30% of the cost of this industry and there is a huge scope for ornamental feeds in India.

Nutrients essential for ornamental fish are the same as those required by most other food fishes like the exclusive requirement of proteins (amino acids), lipids (fats, oils, fatty acids), carbohydrates (sugars, starch), vitamins and minerals. In addition, pigments (carotenoids) are commonly added to the diet of ornamental aquarium fishes to enhance their skin coloration and the primary objective of the feed is to maintain the skin colour and health of the ornamental fish rather than the growth per se. Hence while formulating the feed for aquarium fishes the critical nutrients required for proper colouration and better health are to be taken care of in addition to the maintenance requirement of the fish. The synthetic pigment, astaxanthin and canthaxanthin are the most commonly used additive (100–400 mg/kg). The blue-green algae, Spirulina, dried shrimp meal, crab shell meal, shrimp and palm oils, and extracts from marigold, red peppers and phaffia yeast are excellent natural sources for pigmentation. About 80% of the ornamental fishes in display aquariums are fed with exclusively prepared foods that most commonly are produced in the form of flakes, pellets or tablets. The larval and fry stages of the fish are fed with micro feeds in the form of microparticulated/micro extruded/micro encapsulated/ flakes/ extruded spheres and pellets.

CIBA has done extensive research on the development of feeds for ornamental fishes (Figure 16.6) and developed and branded it as Kalor fish Plus, a versatile feed for all aquarium fishes. This feed has been extensively tested and the technology has been commercialized on a non-exclusive basis. Though there are some Indian companies that recently ventured into aquarium feed manufacturing, the majority of the market share is still with imported feeds. There is a huge demand for quality ornamental feeds to improve larval survival in the hatchery and to improve the health and colour of the aquarium fish are growing. However, imported feeds are very expensive and there is no assurance about the quality of the feeds. ICAR-CIBA realized the urgent need for an indigenous micro feed technology, especially for ornamental feeds, and accomplished it. This will be of great use to people involved in the ornamental aquarium business, aquarium hobbyists and breeders.

Fig. 16.6: Feeds for ornamental fishes

Fish as a health food

The food we consume should maintain a good state of health and optimal performance by providing all the essential nutrients required in adequate quantity and in correct proportions to prevent deficiency disorders and chronic diet related diseases. Fish is mainly considered to be a nutritionally superior and comparatively cheaper food source than terrestrial non-vegetarian meat sources nowadays, which plays a crucial role in global food nutrition security, especially in low and middle-income countries (Reksten et al., 2020). It serves a healthy, safe, nutritious and balanced diet to human beings. Fish is rich in valuable nutrients, especially digestible protein with essential amino acids (EAA) similar to the requirement of humans, contains an adequate amount of long-chain n-3 polyunsaturated fatty acids (LC n-3 PUFA), trace elements etc., (WHO, 2003).

Intake of unsaturated fatty acids is considered better than saturated fatty acids as the latter fatty acids stimulate the body to synthesize more low-density lipoproteins which are 'un health cholesterol'. Fish and shrimp lipids differ greatly from mammalian lipids in that they include up to 40% of long-chain fatty acids (C14-C22) that are highly unsaturated and contain highly unsaturated fatty acids with 5 or 6 double bonds. Further, PUFAs are found in proportions of about 35-55 % in shrimp body lipids. Also, the PUFA belongs to the n-3 or ω-3 groups. There is no other known food source that can provide n-3 PUFA in such abundance. The presence of high levels of the long-chained ω-3 fatty acids, EPA and DHA, is identified as one of the major benefits of ingesting fish and shellfish species derived lipids (Mahaffey, 2004). Interestingly, small amounts of EPA and DHA can reverse an n-3 Fatty acid deficiency. EPA is a substrate for eicosanoid synthesis. DHA, however, is the most abundant n-3 fatty acid present in tissues, which has led to the conclusion that DHA is an essential n-3 fatty acid. This uniqueness will drive the demand for food from aquatic origin. Hence there will be a huge market for designer foods from aquaculture which has immense potential for entrepreneurship.

Conclusion

The Indian aqua feed industry has rapidly evolved over the years to cater to the needs of fast-growing aquaculture during the last decade. However, the specialty feed sector, particularly larval, broodstock, ornamental and functional feeds are continued to depend majorly on imported feed. The establishment of state-of-the-art facilities for specialty feed production coupled with scientific feed formulation and innovative feed management will open many avenues for entrepreneurs and ultimately support the rapidly growing aquaculture industry. This would provide economic and nutritional security to the Indian population through high-quality proteins and micronutrients. The nutritional richness of food of aquatic origin will further promote designer foods and create demand for increased food from aquaculture. There is a huge potential for entrepreneurship in the aqua feed sector in the years to come.

References

Ambasankar, K. Kumaraguru Vasagam, K.P and Vijayan. K.K. 2017. Current status of shrimp feed sector in India. Aqua Aquaria India .97-105

Brander, K.M. 2007. Global fish production and climate change. Proceedings of the National Academy of Sciences 104(50), 19704–19714.

FAO. 2022. The State of World Fisheries and Aquaculture 2022. Towards Blue Transformation. Rome, FAO. https://doi.org/10.4060/cc0461en

FAOSTAT(2019). Food Agriculture and Organization (FAOSTAT). Retrieved from http://www.fao.org/faostat/en/#data/QC

Soto, J.O., de Jesús Paniagua-Michel, J., Lopez, L., & Ochoa, L. 2015. Functional feeds in aquaculture. In Springer Handbook of Marine Biotechnology (pp. 1303– 1319). Berlin Heidelberg: Springer

Mahaffey, K.R., 2004. Fish and shellfish as dietary sources of methylmercury and the ω-3 fatty acids, eicosahexaenoic acid and docosahexaenoic acid: risks and benefits. Environmental research. 95(3): 414-428.

Reksten, A.M., Somasundaram, T., Kjellevold, M., Nordhagen, A., Bøkevoll, A., Pincus, L.M., Rizwan, A.A., Mamun, A., Thilsted, S.H., Htut, T., Aakre, I., 2020. Nutrient composition of 19 fish species from Sri Lanka and potential contribution to food and nutrition security. Journal of Food Composition and Analysis. 91: 103-108.

Soto, J. O., de Jesús Paniagua-Michel, J., Lopez, L., & Ochoa, L. (2015). Functional feeds in aquaculture. In Springer Handbook of Marine Biotechnology (pp. 1303– 1319). Berlin Heidelberg: Springer

World Bank (2010) World Development Report 2010: Development and Climate Change. The World Bank, Washington, DC, USA, 424 pp.

17

Overview and Status of Fisheries Development in The Country: Issues and Challenges

Manas Kumar Sinha

Fishery Survey of India, Sasoon Dock, Colaba, Mumbai-400005, Maharashtra

Abstract

India has bountiful resources comprising inland, brackish and marine water along with diverse fish and shellfish species. Indian Fisheries sector evolved gradually over the years and became an important socio-economic attribute for the nation. India is the third-largest fish and aquaculture-producing country and accounts for about 16% of total inland and 5% of total global marine fish production respectively. In 2021-22, India's total marine and inland fish production stood at 16.24 million metric tonne (MMT), which includes 12.12 MMT and 4.13 MMT from inland and marine sectors, respectively. The annual growth rate in the marine sector was observed at around 18.7 %, whereas inland sector achieved more than 7.76 % during 2021-22. To increase the marine and inland finfish production in the country during May, 2020, the Govt of India announced Pradhan Mantri Matsya Sampada Yojana (PMMSY) as a tool of Atma Nirbhar Bharat towards the purpose of rural development by utilizing rural resources and boosting rural economy in a rapid way. The present paper depicts the overview on the status of fisheries development in the country highlighting the issues in each sector of fisheries. The paper focused more the seed requirement in order to develop the inland aquaculture including mariculture. Finally,the paper draws the way forward for a sustainable fish production in the country.

Keywords: *Aquaculture, Challenges for sustainable fisheries, Livelihood, Nutrition security, Status of fisheries development*

Introduction

Fisheries and Aquaculture remains to be an important source of food, nutrition, income, and livelihood to millions of people. Fish is an affordable and good source of animal protein, a good option to reduce hunger and malnutrition in the country. India has bountiful resources comprising inland, brackish and marine water along with diverse fish and shellfish species. Indian Fisheries sector evolved gradually over the years and became an important socio-economic attribute for the nation. India is the third-largest fish and aquaculture-producing country and accounts for about 16% of total inland and 5% of total global marine fish production respectively. In 2021-22, India's total marine and inland fish production stood at 16.24 million metric tonne (MMT), which includes 12.12 MMT and 4.13 MMT from inland and marine sectors, respectively. The annual growth rate in the marine sector was observed at around 18.7 %, whereas inland sector achieved more than 7.76 % during 2021-22 (Table 17.1).The fishery potential of the country has been estimated to the tune of 22.31 MMT out of which Marine fisheries potential is 5.31 MMT and Inland fishery potential is 17 MMT. (Table 17.3).

Table 17.1: Fish Production in India for the period 2010-11 to 2021-22

Year	Fish production (In Lakh Tonnes)			Annual Average Growth Rate (Percentage)		
	Marine	Inland	Total	Marine	Inland	Total
2010-11	32.50	49.81	82.31	4.70	1.78	2.91
2011-12	33.72	52.94	86.66	3.75	6.28	5.28
2012-13	33.21	57.19	90.40	-1.51	8.03	4.32
2013-14	34.43	61.36	95.79	3.67	7.29	5.96
2014-15	35.69	66.91	102.60	3.66	9.04	7.11
2015-16	36.00	71.62	107.62	0.87	7.04	4.89
2016-17	36.25	78.06	114.31	1.14	8.63	6.12
2017-18	37.56	89.48	127.04	3.61	14.62	11.13
2018-19	38.53	97.20	135.73	2.58	8.62	6.84
2019-20	37.27	104.37	141.64	3.2	7.37	4.35
2020-21	34.76	112.49	147.25	6.7	7.80	4.0
2021-22	41.27	121.21	162.48	18.7	7.76	10.34

Source: Hand Book of Fisheries Statistics, 2022, GOI

Marine fisheries activities are spread along the country's long coastline (8118 km) and in the vast Exclusive Economic Zone (EEZ 2.02 million km^2). The inland resources are in the form of rivers and canals (0.19 million km), floodplain lakes (1.0 million ha), ponds and tanks (2.43 million ha), reservoirs (3.15 million ha), brackish water (1.18 million ha), saline/alkaline affected areas (0.24 million ha), and such others etc., covering about 9 million ha, providing livelihood to 16 million (Table 17.2). In this direction sea cage culture may play an important role in rapid expansion of seafood production at global level. In present scenario, due to limited availability of fish fingerling we may not expand sea cage culture to the optimum level. Therefore, need of hour is to strengthen the available infrastructure and other needed backward and forward support system enhancing for fish seed production apart from developing market linkages with diversification of species.

Fisheries sector plays a crucial role in the national economy and is one of the key contributors to the country's foreign exchange earnings. In 2021-22, 77.72% of the Marine Fisheries and 71.71% of the Inland Fisheries of the respective potential has been harnessed (Table-2). The Annual Average Fish Production in India is consistently increasing over the years from (76.16 lakh tones) 2009-10 to (162.48 lakh tones) 2021-22. Andhra Pradesh, West Bengal, Karnataka, Odisha and Gujarat evolve to be the five major fish-producing states in India.

Table 17.2: Available Fishery Resources of India.

Sl. no	Type of Resources	Area
A. **Inland Fishery Resources**		
1	Length of rivers and canals	0.19 million km
2	Reservoirs	3.15 million ha
3	Ponds and tanks	2.36 million ha
4	Flood plains / derelict water bodies	1.00 million ha
5	Inland saline/ alkaline land	0.23 million ha
6	Brackish water	1.19 million ha
B. **Coldwater Fishery Resources**		
1	Rivers	8253 km
2	Natural lakes	21900 ha
3	Reservoirs	29700 ha
C. **Marine Fishery Resources**		
1	Length of coastline	8118 km
2	Exclusive Economic Zone	2.02 million sq. km
3	Continental shelf	0.53 million sq.km

Source: Hand Book of Fisheries Statistics, 2022, GOI

More than 75% of inland production comes from fish farming i.e from about 70% of the available 2.43 million ha of ponds and tanks resource in the country. Horizontal expansion, even to the 100% utilization level, may not be sufficient to cater to the fish demand in coming days. Though ponds and tanks have remained as the major resources for aquaculture production, the potential of the 0.19 million km rivers and canals, 3.15 million ha reservoirs, and the 0.8 million ha of floodplain lakes and derelict waters in the country is yet to be harnessed. Therefore, while it is necessary to increase the present average productivity from 3.0 t/ha to a level of 5.0 t/ha, holistic development of the other natural waters also needs to be explored to reduce the stress on pond resource and ensure sustainable development. One of the major constraints for the Fresh water aquaculture was the availability of seed in both quantity and quality, particularly of the commercially important carp species which contributes nearly 65 to 85 % to total fresh water fish production, such as the Indian major carps (IMC) , viz., Catla (Labeo catla), Rohu (Labeo rohita) and Mrigal (Cirrhinus mrigala), Exotic carps such as Common carp (Cyprinus carpio), Grass carp (Ctenopharyngodon idella) and Silver carp (Hypophthalmichthys molitrix) etc., at right time, at right place, at affordable cost. Fresh water aquaculture in ponds and tanks is predominantly rain dependent and different parts of the country receive at different times and the seed requirement will be at different times. Hence making available seed throughout the year needs focused attention through innovative programs like early breeding of IMC where ever it is required and networking of seed growers across the country.

Table 17.3: Sector Wise Estimated Fishery Potential of India

Sl.No	Types of Fisheries Resources	Estimated Potential	Percentage Harnessed (Till 2021-22)
1	Marine	05.31 Million Tonnes	77.72 %
2	Inland	17.0 Million Tonnes	71.71%
3	Total	22.31 Million Tonnes	72.83%

Source: Fishery Survey of India, 2018 and ICAR-CIFA, 2022

The major requirements for aquaculture are land, water, good breed, quality seed and feed. Availability of these inputs is necessary to guarantee optimum production from aquaculture. Some of the most pressing issues concerning seed are inadequate and unreliable brood stock of good genetic makeup, supply of quality seed, hatchery technology, not following standard operational procedures (SOP) and inadequate infrastructure facilities for rearing fry/ fingerlings, distribution mechanisms, impacts on release of cultured seed stocks, the need for more hatcheries with business orientation and others. In order to secure stable seed supply, factors affecting seed availability, seed quality, seed production technologies and support services, seed distribution networks, breeding technologies, genetic improvement and domestication need to be understood well. In Indian perspective, the availability of required quantity of seed of desired species at the appropriate time is a prerequisite for rapid expansion, growth and success of aquaculture operation.

Brood stock or brood fish are base materials on which the growth of aquaculture industry depends. Brood fish are parent fish from which fry and fingerlings are produced. The quality and reliable supply of healthy fry and fingerlings having sound genetic base is largely depending on the successful stocking and rearing of brood fish. Brood fish collection, rearing and management are the most important parts of aquaculture activities. Importance of brood fish management is to ensure quality of eggs and sperm; increase fecundity; produce stronger and disease free larvae and fry; spontaneous and timely supply of seeds; removal of the inbreeding problem; endanger fish from extinction; and successful aquaculture with high production potentiality. In order to achieve the targeted production it is essential to fill the gap of availability of quality seed material by increasing the seed production in both the sector. Fish seed Production in India as on 2020-21 is 5,40,690 lakh fry (Handbook on fisheries statistics, 2022).

To increase the marine and inland finfish production in the country during May, 2020, the Govt of India announced Pradhan Mantri Matsya Sampada Yojana (PMMSY) as a tool of Atma Nirbhar Bharat towards the purpose of rural development by utilizing rural resources and boosting rural economy in a rapid way. PMMSY can be a key factor for augmenting livelihood of rural sector of the nation. The main motto of PMMSY is **'Reform, Perform and Transform'** in the fisheries sector.

The reforms and initiatives in PMMSY scheme have been inculcated in core & trunk infrastructure development, modernization of Indian fisheries, especially the push for new fishing harbors/landing centers, modernization and mechanization of traditional fishermen crafts-trawlers-deep sea going vessels, provision of post-harvest facilities to reduce post-harvest loss, cold chains facilities, clean and hygienic fish markets, two wheelers with ice boxes and many more. Fishermen are provided with insurance cover, financial assistance and a facility of Kisan Credit Card as well. PMMSY has set a way forward to enhance its utilization by invigorating rural economy.

Inland Aquaculture Status

Aquaculture is being the fastest food production sector in the world with an estimated production of about 179 million tonnes. The fish production from capture fisheries (96 million tonnes) is almost stagnant for many decades but the demand is rising significantly due to the awareness about the health benefits of fish consumption. The additional demand for fish consumption must be achieved only through aquaculture. The current fish production from aquaculture has increased from 44 million tonnes (2005) to 82 million tonnes

(2018) and it is sharing almost 45% of total global fish production but this production level is not sufficient to fulfil the demand (FAO, 2020). This article focuses on the present status and future scope of freshwater aquaculture in India to achieve blue revolution in India

Status of inland Fish Production in India

India ranks second in world inland fish production, next to China (FAO, 2020). During last seven decades, Indian fisheries have grown more than 21 folds with an increase in fish production from 0.75 MMT in 1950-51 to 16.24 MMT in 2021-2022 (Hand book on Fisheries Statistics, 2022). Since the last three decades, freshwater aquaculture has also been compensating the loss in the fish production due to decline in the capture fisheries as well as increasing overall fish availability in the country. In past five year, the Inland production is increased from 8.93 MMT to 12.12 MMT (Table 17. 4 and Figure 17.1). In the year 2021-22, the top leading five inland fish production states are Andhra Pradesh (0.421 MMT), West Bengal (0.165 MMT), Uttar Pradesh. (0. 081 MMT), Odisha (0.789 MMT) and Bihar (7.62 Lakh tons). The state of Andhra Pradesh always leads the Inland fish production. (Table 17.4 and Figure 17.2)

Table 17.4: State Wise Fish Production in India During 2017-18 to 2021-22 (Production in Lakh Tonnes)

	Years	2017-18		2018-19		2019-20		2020-21		2021-22	
Sl. No	**State/UT's**	**Inland**	**Marine**	**Inland**	**Marine**	**Inland**	**Marine**	**Inland**	**Marine**	**Inland**	**Marine**
1	Andhra Pradesh	28.45	6.05	33.91	6.00	36.1	5.64	40.7	5.54	42.19	5.94
2	Arunachal Pradesh	0.04	0.00	0.05	0.00	0.05	0.00	0.05	0.00	0.05	0.00
3	Assam	3.27	0.00	3.31	0.00	3.73	0.00	3.93	0.00	4.17	0.00
4	Bihar	5.88	0.00	6.02	0.00	6.41	0.00	6.83	0.00	7.62	0.00
5	Chhattisgarh	4.57	0.00	4.89	0.00	5.72	0.00	5.77	0.00	5.91	0.00
6	Goa	0.06	1.18	0.05	1.15	0.04	1.01	0.05	1.06	0.05	1.11
7	Gujarat	1.38	7.01	1.42	6.99	1.58	7.01	1.57	6.83	1.86	6.88
8	Haryana	1.9	0.00	1.8	0.00	1.91	0.00	2.03	0.00	2.08	0.00
9	Himachal Pradesh	0.13	0.00	0.13	0.00	0.14	0.00	0.15	0.00	0.16	0.00
10	Jharkhand	1.9	4.14	2.08	0.00	2.23	0.00	2.38	0.00	2.57	0.00
11	Karnataka	1.88	4.1284	1.98	3.9	2.29	4.03	2.61	3.47	4.85	5.89
12	Kerala	1.89	0.00	1.92	6.09	2.05	4.75	2.24	3.92	2.25	6.01
13	Madhya Pradesh	1.43	4.75	1.73	0.00	2.00	0.00	2.49	0.00	2.93	0.00
14	Maharashtra	1.31	0.00	1.00	4.68	1.18	4.43	1.25	3.99	1.57	4.33
15	Manipur	0.33	0.00	0.32	0.00	0.32	0.00	0.33	0.00	0.33	0.00
16	Meghalaya	0.12	0.00	0.13	0.00	0.14	0.00	0.16	0.00	0.18	0.00
17	Mizoram	0.08	1.51	0.07	0.00	0.07	0.00	0.05	0.00	0.05	0.00
18	Nagaland	0.09	0.00	0.09	0.00	0.09	0.00	0.09	0.00	0.09	0.00
19	Odisha	5.34	6.05	6.00	1.59	6.6	1.58	7.01	1.72	7.89	2.01
20	Punjab	1.37	0.00	1.35	0.00	1.51	0.00	1.65	0.00	1.90	0.00

	Years	2017-18		2018-19		2019-20		2020-21		2021-22	
Sl. No	**State/UT's**	**Inland**	**Marine**	**Inland**	**Marine**	**Inland**	**Marine**	**Inland**	**Marine**	**Inland**	**Marine**
21	Rajasthan	0.54	0.00	0.55	0.00	1.16	0.00	0.6	0.00	0.66	0.00
22	Sikkim	0.00	0.00	0.00	0.00	0.00	0.00	0.00	0.00	0.00	0.00
23	Tamil Nadu	1.85	4.97	1.7	5.20	1.74	5.83	1.75	5.48	2.12	5.95
24	Telangana	2.70	0.00	2.84	0.00	3.00	0.00	3.49	0.00	3.90	0.00
25	Tripura	0.77	0.00	0.70	0.00	0.78	0.00	0.82	0.00	0.82	0.00
26	Uttar Pradesh	6.29	0.00	6.62	0.00	6.99	0.00	7.46	0.00	8.09	0.00
27	Uttarakhand	0.05	0.00	0.05	0.00	0.05	0.00	0.06	0.00	0.06	0.00
28	West Bengal	15.57	1.85	16.19	1.63	16.19	1.63	16.69	1.55	16.52	1.91
29	A and N Islands	0.00	0.39	0.00	0.4	0.00	0.4	0.00	0.43	0.00	0.44
30	Chandigarh	0.00	0.00	0.01	0.00	0.01	0.00	0.00	0.00	0.00	0.00
31	Daman and Diu Daman Nagar Haveli	0.00	0.24	0.00	0.28	0.00	0.32	0.00	0.3	0.00	0.30
32	Delhi	0.01	0.00	0.00	0.00	0.01	0.00	0.01	0.00	0.01	0.00
33	Jammu & Kashmir	0.21	0.00	0.21	0.00	0.21	0.00	0.21	0.00	0.25	0.00
34	Ladakh	0.00	0.00	0.00	0.00	0.00	0.00	0.00	0.00	0.01	0.00
35	Lakshadweep	0.00	0.21	0.00	0.22	0.00	0.2	0.00	0.15	0.00	0.12
36	Puducherry	0.07	0.42	0.07	0.4	0.07	0.44	0.05	0.34	0.08	0.39
	INDIA TOTAL	89.48	37.56	97.20	38.53	104.37	37.27	112.48	34.77	121.21	41.27

Source: Hand Book of Fisheries Statistics, 2022, GOI

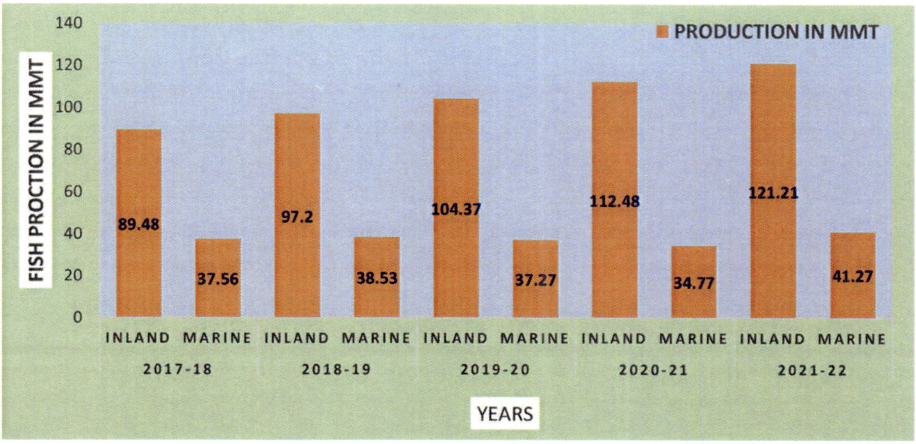

Fig. 17.1: Year Wise Fish Production in India During 2017-18 to 2021-22

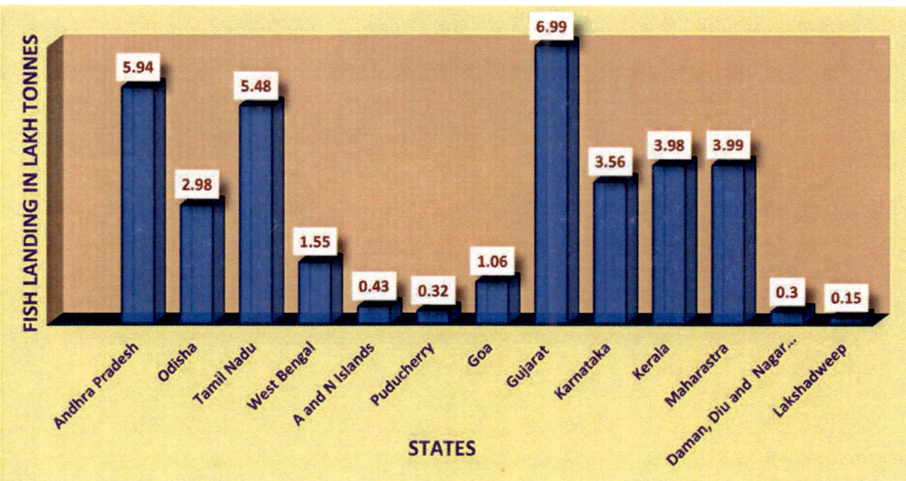

Fig. 17.2: State-Wise Marine Fish Landing in India During 2021-22

Species diversification in Freshwater Aquaculture

India is blessed with rich fauna and flora diversity. Our country has diversified fish species and only little amount of it is utilized for farming activities. Besides of Indian major carps (catla, rohu, mrigal), our country also possesses several other cultivable minor carps and barbs such as, *Labeo calbasu, L. fimbriatus, L. gonius, L. bata, Osteobrama belangeri and Puntius sarana, P. gonionotus* etc., Likewise few culturablee potential catfishes such as, magur *(Clarias batracus)*, butter catfish (*Ompok pabda*), singhi (*Heteropneustes fossilis*), striped catfish (*Pangasionodon hypophthalmus*), striped murrel (*Channa striata*), climbing perch (*Anabas testudineus*), scampi (*Macrobrachium rosenbergii*), tengra

(*Mystus cavasius*), *yellow catfish (M. gulio,)* mola, freshwater eels, loaches are also commercially farmed diversified fishes in India. Indian major carps have been the major species of interests in freshwater aquaculture. Catla, rohu and mrigal are the important carp species that are cultured traditionally in ponds, cages, pens, etc. Owing to the compatibility, with regards to food and habitat preference, these three species are cultured together in ponds under polyculture system. Figure 17.3 indicates that the Major carp contributes 57 % of the Indian aquaculture production followed by others (19%) , Exotic carps (13%) , Cat fishes and Minor carps (5% each) and Murrel (1%). Generally, in major carp polyculture, 6000- 8000 fingerlings/ha are stocked for a production target of 4-6 tonnes/ha/yr.

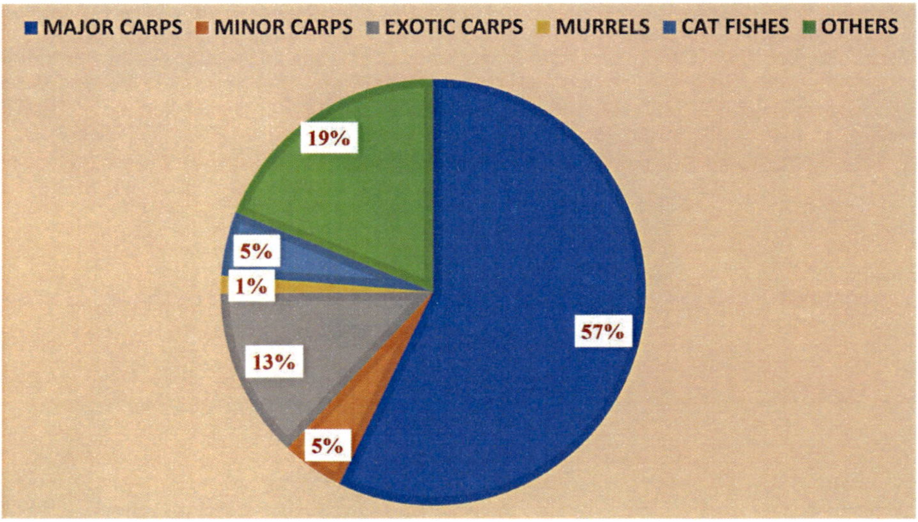

Fig. 17.3: Percentage of Species Composition During 2020-21
Source: Hand Book of Fisheries Statistics, 2022, GOI

Status of Brackish Water Aquaculture

India has 1186442.40 ha of land area that has potential for brackish water aquaculture, however only 184263.43 ha (15.5%) area are used for brackish water aquaculture. Apart from that India has 233900 ha of saline/ alkaline soils, out of which only 147 ha (0.063%) is underutilization (Table 17.5). The cumulative brackish water area available in India now stands at 14.2 Lakh hectares out of which only 1.84 lakh ha (12.98 %) are under culture. West Bengal and Gujarat have higher potential for brackish water aquaculture owing to the high tidal amplitude. Andhra Pradesh developed almost 42.8 percent of available area for shrimp culture, whereas Maharashtra and Gujarat utilized only 21.4 and 2.05 percent respectively of the available area. In recent days the Saline/ Alkaline Soils available at Haryana, Punjab, Rajasthan and Uttar Pradesh are in the news for its culture potential. In addition to this there are about 3.9 million ha of estuaries and 0.5 million ha coastal mangrove areas available in the Country which also needs to be utilized. Brackish water estuaries are another set of water bodies which has huge potential for both fish and shellfish culture. Shrimps, oysters, mussels, crabs, lobsters, sea bass, groupers, mullets, milk fish, cobia, silver pompano, pearl spot, ornamental fishes and weeds are being cultivated in the brackish waters.

A major shift in policy on shrimp farming took place with the introduction of an exotic species of shrimp, namely, *Litopenaeus vannamei* and technologies for breeding and seed production of Barramundi or Sea bass, (*Lates calcarifer*), crabs (*Scylla serrata* and *Scylla tranquebarica*), milk-fish (*Chanos chanos*), pearls pot (*Etroplus suratensis*) and mullets (*Mugil* spp.). The Coastal Aquaculture Authority (CAA) is regulating these activities in saline and brackish water systems within 2 km from the High Tide Line for sustainable development of coastal aquaculture sector.

The traditional brackish water aquaculture practices in India include bheries (manmade impoundments in coastal wetlands) in West Bengal and pokkali cultivation (salt resistant deep-water paddy) in Kerala. In the traditional system of culture, tidal water is impounded in the inter-tidal mudflats by raising bunds and the marketable sized fish and shrimps are harvested during spring. These systems have been sustaining production levels between 500-750 kg/ha/year with shrimps contributing 20-25 percent of the total production. However, from one hectare area easily 8000 Kg can be harvested if scientific methods are adopted in culture practices. This shows that, there is vast opportunity for the development of brackish water fish/ shell fish farming in India. Thus against the total 113.6 Lakh tonnes is possible potential of from 1.42 Million ha at the rate of 8 tonnes/ha and additional 0.06 tonnes/ha from estuaries and Mangroves, thereby a total of 113.852 Lakh tonnes from a cumulative total of 5.82 million ha. Whereas during 2019 the total shrimp production in India has reached 8,04,00 tonnes. Andhra Pradesh, with 71 per cent share, is the largest producer of shrimp in the country. West Bengal (10%), Odisha (9%), Gujarat (5.5%) and Tamil Nadu (2.7%) are the other producers. Lower salinity in the top three states has led to better crops. More water salinity in Gujarat and Tamil Nadu hit production of shrimps.

Brackish water aquaculture in the country has expanded steadily from a traditional activity to reach the present commercial scale. The growth of shrimp aquaculture sector was phenomenal during the 90's and this has resulted in the all-round development in shrimp hatcheries, feed mills, ancillary industries like ice plants, processing plants, drugs and chemical land other aquaculture related engineering products. The sector has generated more than **three lakhs'** jobs. During 2020-21 India has been producing nearly 8.5 Lakh tonnes of shrimp (Figure-4). The production could have gone beyond 1 million tonnes but due severe disease problem the shrimp production though increasing but with a very slow pace.

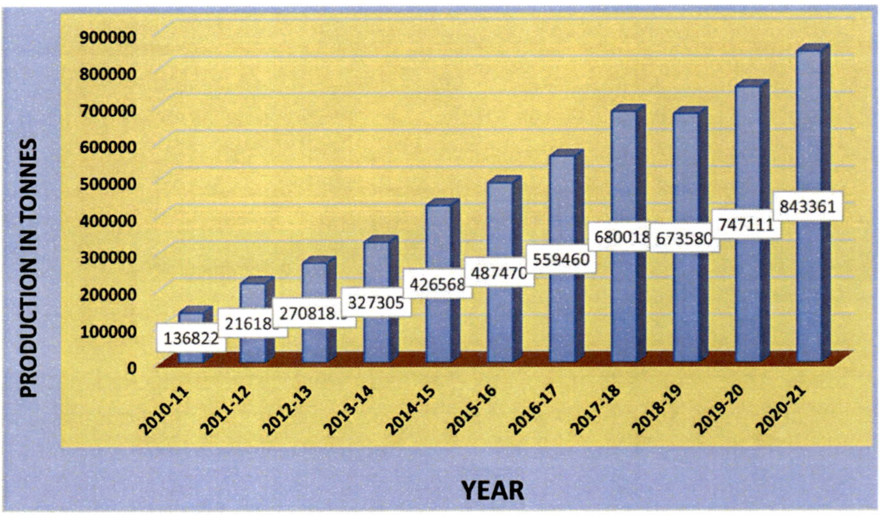

Fig. 17.4: Shrimp Production in India during 2010-11 to 2020-21.

Regarding Seabass culture, the seed production technology has been standardized by the tireless efforts of Central Institute of Brackish water Aquaculture, Chennai. The same is now being propagated by Rajiv Gandhi Center for Aquaculture, a unit of MPEDA, Cochin. But, chain of hatcheries is yet to be set up in order to cater the demand for seed by the farmers and entrepreneurs. The culture of sea bass in brackish water pond is fairly standardized. Since sea bass is a predatory fish, it can be cultured with tilapia which will reduce the feed cost and subsequently cost of production. In case of mullets, milkfish although culture technology has been perfected, the seed production technology has not been developed. Hence the aqua culturists are still dependent upon wilds seed collection.

The scope of culture of mud crab is enormous throughout the coastline of India. However the seed production technology is still in a very nebulous stage. Though CIBA and Centre for Marine Living Resources, Andhra University has standardized the hatchery technology of mud crab but still a lot need to be done for commercial seed production of Mud crab. The culture technology of mud Crab, *Scylla serrata* has been standardized in the tide-fed brackish water ponds. The scope of mud crab fattening is also substantial. Entrepreneurs are earning a good return by fattening the big size soft crabs in pen type culture. Once the shell is hardened, it is fetching a good price at market. Few BW aqua farmers have taken up mud crab culture but the major hurdle is seed and the feed for mud crabs.

The State like Haryana, where ground water has become saline, with slight modification of technique tiger shrimp has been cultured as demonstrated by CIFE Centre at Haryana. This shows an enormous possibility of brackish water aquaculture in inland areas. Hence, it can be said that once the seed production technology of commercially important brackish water fishes are standardized, brackish water aquaculture is all set to spread all over the country.

Table 17.5: State wise area availability of Brackish-water area and area under Coastal Aquaculture.

Name of the States/Union Territories	Total Brackish water Area as reported by States/UT's (in hectares)	Area under brackish water aquaculture (in hectares)
A. Brackish Water		
Andaman & Nicobar Islands	248.00	0.00
Andhra Pradesh	150000.00	64222.00
Goa	18500.00	131.68
Gujarat	370000.00	7597.00
Karnataka	8000.00	4000.00
Kerala	65213.00	13042.00
Maharashtra	80000.00	17125.15
Odisha	32587.00	15308.00
Puducherry	894.40	95.28
West Bengal	405000.00	59000.00
Tamil Nadu	56000.00	3742.32
Total Brackish water	1186442.40	184263.43
B. Saline/ Alkaline Soils		
Rajasthan	70000.00	0.00
Punjab	7000.00	100.00
Haryana	20000.00	40.00
Uttar Pradesh	136900.00	7.0
Total Saline/ Alkaline Soils	233900.00	147.00
Grand Total (A+B)	1420342.40 (14.2 lakh ha)	184410.43 (1.84 lakh)

Source: MPEDA

Status of Marine Fisheries

The marine fisheries sector contributes more than 40% of its total fish production in India. The sector is also supplying protein-rich food to the country's population at less cost, and generates lucrative foreign exchange earnings through seafood export. Scientific stewardship of our marine resources entails constant vigil in maintaining the biodiversity and ecosystem balance. Strategies for the sustainable management of marine fishery resources are developed through the timely collection and analysis of scientific data on marine fisheries, especially the production aspect. (FRAEED, CMFRI, 2023)

During 2022, the estimated marine fish landings along the coast of India (mainland) was 3.49 million tonnes, indicating 14.53 % increase of compared to the landings in 2021. Among the coastal states, Tamil Nadu stood in the first with 7.22 lakh tonnes followed by Karnataka with 6.95 lakh tonnes and Kerala with 6.87 lakh tonnes in fish landings. The Gujarat state, which had been occupying the top two slots during the previous years, appropriated the fourth with 5.03 lakh tonnes in fish landings. The leading four states viz Tamil Nadu, Karnataka, Kerala and Gujarat, contributed with 20.69%, 19.90%, 19.68%, and 14.40% of the total fish landing respectively. Except for Odisha and Gujarat, all states witnessed an increase in landings compared to 2021. The observed increase was 38.42% in West Bengal landings, 4.55% in Andhra Pradesh, 28.38% in Tamil Nadu, 30.44% in Puducherry, 23.76% in Kerala, 26.13% in Karnataka, 1.22% in Goa, 38.19% in Maharashtra and 10.38% in Daman & Diu. Landings in Odisha and Gujarat witnessed a reduction to the tune of 26.07% and 12.79%, respectively in 2022 compared to 2021. Compared to 2019 the state of Goa (58.94%), Karnataka (38.57%), Puducherry (35.34%), Odisha (30.34%), and Kerala (26.29%) observed significant jumps in the estimated fish landings. (FRAEED, CMFRI, 2023)

The species wise marine fish landing during 2022 indicates that Indian mackerel(3.28 lakh tonnes) dominated the catch followed by Oil sardine (2.51 lakh tonnes), Ribbon fishes (2.27 lakh tonnes), Cephalopods (2.06 lakh tonnes), Threadfin breams (1.99 lakh tonnes), Lesser sardines (1.69 lakh tonnes), Scads (1.63 lakh tonnes), Penaeid prawns (1.63 lakh tonnes), Anchovies (1.57 lakh tonnes) and Non-penaeid prawns (1.47 lakh tonnes). The year 2022 had been a comeback year of sorts for the Oil sardine (*Sardinella longiceps*), one of the most preferred marine fishes in India. The landings of the same witnessed an increase of 188.15% in 2022 compared to 2021 and secured second position among the marine resources in quantity landed. (FRAEED, CMFRI, 2023)

The different crafts which are used for fishing indicates that mechanized fishing vessels landed 82 % of catch (2.85 million tonnes followed by motorized fishing crafts with 17 % (0.61 million tonnes) and non-motorized fishing crafts 1% (0.04 million tonnes) of the total landings. The Pelagic resources contributed 57 % (1.98 million tonnes) of the total landing with the major contribution from Indian mackerel, Oil sardine and Ribbon fishes followed by Demersal resources with 26% (0.91 million tonnes) , Crustacea with 11 % (0.37 million tonnes) and Mollusca with 6 % (0.22 million tonnes).

Status of Fish Consumption and Disposition

Fish has been an important dietary source of protein and other nutrients throughout human history. Fish can be prepared in a variety of ways. It can be served uncooked (raw food, e.g., sashimi); cured by marinating (e.g., ceviche), pickling (e.g., pickled herring) or smoking (e.g., smoked salmon); or cooked by baking, frying (e.g., fish and chips), grilling, poaching or steaming. Many of the preservation techniques used in different cultures have since become unnecessary but are still performed for their resulting taste and texture when consumed.

Disposition of Fish Catch covers activities such as Marketing Fresh, Frozen, Curing, Canning, Reduction etc. Marketing of the Fresh Fish is the highest practiced activity. India has been facing a constant growth in disposition of Fish catch during 2013-2021. Andhra Pradesh, West Bengal, Odisha inter alia states have highest disposition of Fish Catch in 2020-21. Total Fish consumption is highest in states such as Uttar Pradesh, Odisha, Bihar, Kerala, Tamil Nadu etc. and per capita consumption of fish and prawn is highest in Lakshadweep, Kerala respectively. Tripura comes out to be the state with highest no. of households (rural/ urban) consuming fishes. Figure 17.5 depicts the per capita consumption (> 10 kg)of Fish in leading states of India.

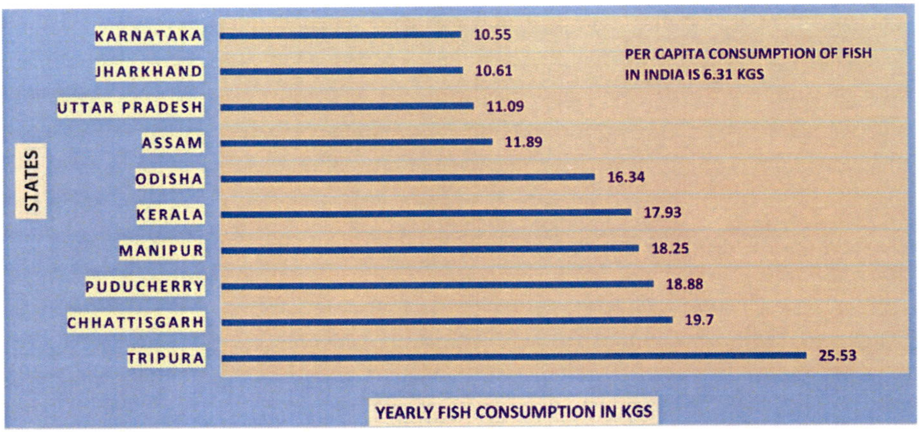

Fig. 17.5: Per Capita Consumption of Fish in Leading States of India

India is among the top five fish exporting countries in the world and at third place preluded by China & Indonesia which constitutes about 1.29% contribution is in Fisheries sector to the Indian economy and about 6.72% contribution is in agriculture. In 2021-22, the country exported 1.36 million MT of seafood worth US $ 7.76 billion (Rupees 57586.48 Crores) which is all time high export by value (Figure-17. 6). India has exported marine products to 123 countries. USA and China are the major importers of Indian seafood. India mainly exports frozen shrimps, fish, cuttlefish, squids, and dried items, live and chilled items. Out of these, frozen shrimp is the major exported marine product contributing to more than 53.18% of the total quantity and about 75.11% of the total export value. In 2021-22, the frozen fish, cuttlefish and squid contributed 6%, 3% and 5% of the total marine products export value respectively.

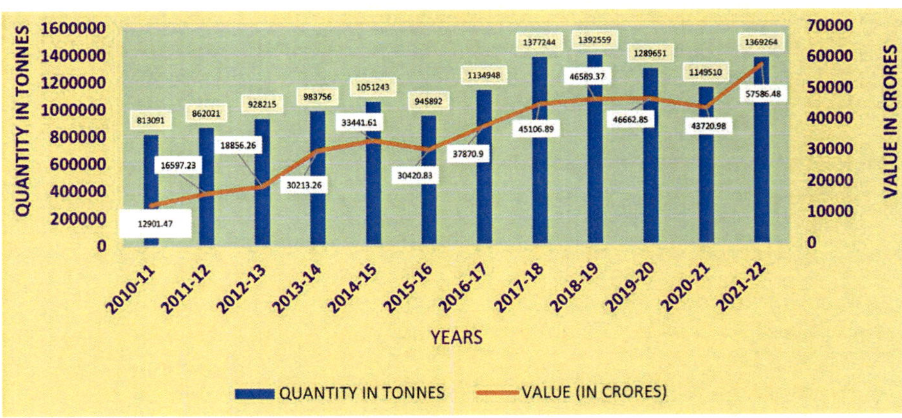

Fig. 17.6: Export of Fish and Fishery Products in India During 2010-11 to 2021-22

Status of Fish Seed Production in India

Fresh Water Fish hatcheries in both the public and private sectors have contributed substantially towards the increase in seed production from 35743 million fry in 2016-17 to over 54068 million fry in 2020-21 (Table-17.6) . Among the states, West Bengal, (12455 million Fry), Jharkhand (11495 million Fry), Assam, (9886 million Fry) Uttar Pradesh (3233 million Fry) and Chhattisgarh (2880 million Fry) are the top five major seed producing states in India. The top 10 states contributes around 90% to the total seed production (Table 17.7) in the country. The seeds reflected in the table are predominantly of Carps as Indias fish production mainly revolves around carps (nearly 90%). Other species seed though are produced but are in very limited in numbers.

Table 17.6: Year Wise Fish Seed Production in India

Sl.No	Year	Fry Production (In Lakhs)
1	2011-12	3,65,651
2	2012-13	3,49,202
3	2013-14	4,14,484
4	2014-15	3,93,487
5	2015-16	3,54,350
6	2016-17	3,57,439
7	2017-18	4,44,207
8	2018-19	4,82,039
9	2019-20	5,21,866
10	2020-21	5.40,690

Table 17.7: State Wise Fish Seed Production in India (Lakh Fry)

State/Union Territory	2016-17	2017-18	2018-19	2019-20	2020-21	Total
Andhra Pradesh	7923	8845	8842	5358	38126	69094
Arunachal Pradesh	65	70	0	0	80	215
Assam	67580	80000	98930	95190	98860	440560
Bihar	7932.2	9206	9286.5	13727.18	17069.4	57221.28
Chhattisgarh	19750.4	22186	25131.8	27386.8	28805.3	123260.3
Goa	6	7	6	3	2.5	24.5
Gujarat	5529.3	1852.3	6340.5	26083.7	9585	49390.8
Haryana	7046	4776.8	4229.6	4538.06	7784.6	28375.06
Himachal Pradesh	625	47	18.3	6.8	15.6	712.7
Jharkhand	41501	103349.1	103629	105692 4	114950	469121.5
Karnataka	4998.8	2481.1	6298	5816.5	7545.6	27140
Kerala	1770.9	1730.4	944.42	1282	1342	7069.72
Madhya Pradesh	11113	11172.3	13729.83	14592.09	14837.5	65444.72
Maharashtra	1895.5	6250	2810.87	13662.7	1964.04	26583.11
Manipur	2150	2500	4060.8	2494.8	2788	13993.6
Meghalaya	75.6	97	65.23	32.14	137.9	407.87
Mizoram	310	426	406.6	400	200	1742.6
Nagaland	480	481	760	795	798	3314
Odisha	7782.8	8077.4	8962.4	9471.6	8441.3	42735.5
Punjab	2375	2779.8	2977.4	2642.35	2562.57	13337.12
Rajasthan	10988.6	10653	10329.3	12264.1	10870.9	55105.9
Sikkim	23.2	6.5	16.25	16.25	14.05	76.25
Tamil Nadu	5732	5572.8	13008.7	19426.88	6032	49772.38
Telangana	1876	4179	6500	5500	6720	24775
Tripura	3137	4350	0	0	0	7487
Uttarakhand	746.5	678.8	655.6	289.6	299.85	2670.35

State/Union Territory	2016-17	2017-18	2018-19	2019-20	2020-21	Total
Uttar Pradesh	27535.6	27128.7	29342.1	30287.8	32331.2	146625.4
West Bengal	115700	124550	124550	124550	124550	124550
A & N Islands	12.8	14.7	0	0	17	44.5
Chandigarh	8	8	0.8	0.8	0.8	18.4
Daman, Diu, Dadra & Nagar Haveli	0	0	0	0	0	0
Delhi	0	20	2	15	0	37
Jammu & Kashmir	770	712	206.6	209.7	457	2354.3
Ladakh	0	0	0	0	0.01	0.01
Lakshadweep	0	0	0	0	0	0
Puducherry	0	0	0	0	0	0
Total	357439.2	444207.7	482039.6	521865.5	540689.82	1852959.9

Source: Hand Book of Fisheries Statistics, 2022, GOI

Need for Improved Fish variety and domesticated Fish species

Quality fish seed is one of the key inputs for augmenting fish production. With use of good quality seed, aquaculture production can only be enhanced as much as 15-30 % without any other additional input or cost. This production can be further enhanced through use of Genetically Improved Fish Seed. Genetically Improved Fish Seed can be reared with existing endowment of farmers and can be harvested obtaining higher yield per unit area. It will not affect or have any advesrse impact on the other indigenous species under culture. The adoption of genetically improved strains will have faster growth rate leading to increase fish production, enhance profitability, and economic gain for the fish farmers as well as the nation. Some of the Gentically Improved Fish Strains also have the disease resistant characterstics which can be a added advantage for the farmers who undertakes commercial intensive culture practices.

Institute like ICAR-CIFA (Jayanti rohu, Improved catla and GI Scampi) ICAR- CIFE (GI Magur/ Maha Magur) and FRIC- Karnataka Veterinary University (Amur Common carp) have developed few Genetically Improved Fish Strains which are superior to the normal strain of the same species. Few Species such as Sea bass, Indian Pompano, Cobia, Tiger Shrimp, Indian White shrimp have also been collected from the nature and Quality Brood stocks have been developed in captivity by ICAR-CIBA,MPEDA, ICAR-CMFRI etc. These captive developed species are also in demand due to standardization of seed production technology, development of quality feed and its culture practices in Pond and cages as the case may be.

Inland Seed Requirements

The total fresh water fish production of the country during 2021-22 was around 12.12 MMT. The projected freshwater fish production for 2024-25 is about 18 MMT. In order to meet the targeted fish production of 18 MMT, it is calculated that the fish seed (All the FW fish species) required is about 2000 crore fingerlings (80-100 mm) and the total spawn requirement will be around 11765 crores (@ 17 % Survival from Spawn to Fingerlings). In order to produce 11764740 million (say 11765 Crore) spawn a total of 23,52,940 Kg brood fish (Male and Female) are required. If a single brood fish weighs 1.0 Kg on an average, we require 23,52,940 nos of brood fishes with an average fecundity of 1,00,000 per female. With a survival of 90% Fingerling attains 1 Kg fish then we require about 26, 14, 380 nos of Fingerlings/ Yearlings to develop as Brooder. If each hatchery can produce 6 crore (60 million) spawn every year then we require 1961 hatcheries in order to produce 11765 crore spawn per annum to achieve 18 MMT fish production in the country. Though the Fecundity, hatching rate and survival varies from species to species but as the Indian major carp constitutes about 90 % of the total production, the assumptions and the projection are mostly based on the carps.

Table 17.9 indicates the species wise break-up of the seed requirement. As the inland FW aquaculture constitutes 70%, we may require 1400 Crores Fingerlings of carps. The other species like Pangassius which needs to be promoted @ 5 % initially may require 100 crores of Fingerlings and Tilapia @10% may require 200 crores of fingerlings. As the Scampi is regaining its importance we may require 125 crores of PL for a targeted production of 50,000 tons under poly culture practices.

Similarly, India has over 600 Shrimp hatcheries (Capacity-125 Billion) along the coastline of India and these hatcheries produced 71 Billion seed in the year 2019. The expected production from shrimp culture in India is about 1.0 MMT. As there is a capacity of 125 billion seed, we may not require further establishment of hatchery particularly in shrimp industry.

Species wise projected seed requirement

As per ICAR-CIFA strategic document submitted to DOF, GOI, at present India have 2405 hatcheries in 27 states. If all the hatcheries are functional then in order to produce 11765 Crore spawn, then each hatchery has target to produce 4.89 crores of spawn (Table- 10). However it needs to be confirmed that how much hatcheries are really in functional stage. From the table-3 it is projected by ICAR-CIFA that at in order to produce 15.03 million metric tons of Inland fish production by 2025 from the present production of 8.89 million tons (2018-19), the additional requirement of hatcheries will be 1305 nos. (assuming 50 million spawn per hatchery). The actionable strategic document FW fish production also indicates that about 3648 tons of carps (equal to 36, 48,000 nos of brood fish of 1 Kg size) is required to produce 182177 crore spawn for production of 15.03 million Inland fish production. (Table-11)

Swain *et al*, 2021 in their document has also projected the above figure assuming 50,000 spawn production per Kg brood fish. As per the assumption of 70 % of Fingerling to brood fish then we require about 52,11,428 nos of Fingerlings as breeder seed which is almost double the nos of breeder seed required as projected above. This is mainly because the fecundity projected is 50 % less in this case.

In order to boost fish production from the same culturable area we require quality brood and seed other than Feed. Therefore, it was very imperative that a projection need to be in place. In this paper the Table-17.9, Table-17.10 and Table-17.11 is given which will certainly give the planners and policy makers to think and act in this direction.

Similarly, the FRP Portable hatchery system developed by ICAR-CIFA for carps, cat fishes including Pabda has also lot of impact in the rural and difficult terrain fish farming area. The developed hatchery can be transported, installed and operated in remote places to ensure easy and timely availability of carp seeds. But however, it is not for higher seed production as in this FRP hatchery only 10-12 kg of carps can be undertaken in one operation. According Swain and Ferosekhan, 2022, at present 27 states have installed these hatcheries and around 210 users were direct beneficiaries of the technologies. FRP Magur hatchery is a very simple device comprising a stand on which are placed a row of plastic tubs (12 cm dia, 6 cm high). Water supplied from the overhead tank through a common pipe to all the tubs with individual control tabs. It includes egg incubation and hatching. The technology creates a suitable environment for high hatching percentage where maximum 50,000 eggs can be incubated at a time. (Swain and Ferosekhan, 2022)

Mariculture Seed Requirement

The current marine fish production in India is about 4.12 million tonnes of which 2 lakh tones comprise of high-value fish that are in great demand for human consumption. The projected marine fish production by 2050 is expected to be a maximum of 6 million tonnes. In view of the vast marine resources available in the country, Sea Cage Farming has emerged as a viable option to increase production of high value marine fish and shellfish. Sea Cage Farming provides immense opportunities for enhancing marine fish production while simultaneously reducing fishing pressure on natural stocks and providing livelihood to coastal communities.

GOI has targeted to install 20000 cages for Open Sea Cage Culture (Mariculture) by the end of 2025, aiming for total production of 35532 tonnes projected production @ 2.5 tonnes/cage from 6.0 m dia. As of now 1,845 cages are installed through ICAR & States Govts. The major species focused under open sea culture are Cobia, Pompano, Seabass, Grouper, snapper. The detailed calculation of the Estimated cages and its requirements are projected in Table 17.12.

The most vital issue for the expansion of the sea cage farming in India is the shortage of fish seeds. Presently seed of Cobia, Pompano, Seabass and Groupers are being produced in a few hatcheries in the country and mainly by the Govt sector hatcheries including research institutes. To meet the growing demand from farmers and enterprenuers, there is an urgent need to produce sufficient quantity of seed either through commercial hatchery production or we need to import with due permission from Govt. till we achieve self-sufficiency in these marine fish seed production.

NFDB has assisted ICAR-CMFRI for standardizing seed production package species such as Cobia, Pompano and groupers and Seabass through ICAR-CIBA. Breeding technologies of these commercialized species have been developed and commercialized by the research institutes, but limited availability of seed is the major constraint for slow growth of mariculture sector in the country. These require the intervention of Govt for infrastructure development for seed production such as specialized hatcheries, brood banks, rearing units etc.

Table 17.9: Species wise break up of Calculated seed requirement

<table>
<tr><th>Species Composition</th><th>Anticipated Percentage of seed required</th><th>Quantity of Fish Seed Required (In crores)</th><th>Stages of Fish seed Required</th></tr>
<tr><td>Jayanti Rohu (Labeo rohita)</td><td rowspan="6">70%
(As per the total Inland Fish production)</td><td rowspan="6">1400 crores</td><td rowspan="6">Fingerlings</td></tr>
<tr><td>Improved catla (Labeo catla)</td></tr>
<tr><td>Amur Common carp (Cyprinus carpio)</td></tr>
<tr><td>Java Putty/ Japanese Putty (Puntius gonionotus)</td></tr>
<tr><td>Pengba (Osteobrama belangiri)</td></tr>
<tr><td>Bata fish (Labeo bata)</td></tr>
<tr><td>Pangassius (P. hypophthalamus)</td><td>5 %</td><td>100 crores</td><td>Fingerlings</td></tr>
<tr><td>Tilapia (Oreochromis niloticus)</td><td>10 %</td><td>200 crores</td><td>Fingerlings</td></tr>
<tr><td>Scampi (Macrobrachium rosenbergii)</td><td>To achieve 50,000 tons of production</td><td>125 crores</td><td>PL</td></tr>
<tr><td>Tiger Shrimp (Penaeus monodon)</td><td rowspan="3">10 %</td><td rowspan="3">1250 Crores
(As per the installed capacity)</td><td rowspan="3">PL</td></tr>
<tr><td>White Shrimp (P. indicus)</td></tr>
<tr><td>Pacific white Shrimp (Litopaenaus vannamei)</td></tr>
<tr><td>Cobia (Rochycentron canadum)</td><td rowspan="5">4%</td><td rowspan="5">80 Crores</td><td rowspan="5">Fingerlings</td></tr>
<tr><td>Silver Pompano (Trachinotus blochii)</td></tr>
<tr><td>Indian Pompano (Trachinotus mookalee)</td></tr>
<tr><td>Sea bass /Barramundi(Lates calcarifer)</td></tr>
<tr><td>Groupers</td></tr>
<tr><td>Any other variety of commercially important</td><td>1%</td><td>2 Crores</td><td>Fingerlings</td></tr>
</table>

Table 17.10: State Wise Existing and Additional Facilities / Infrastructures Requirements and Seed/Brood Projections

Sl.No	Name of the State	No of Carp Hatcheries		Requirements In Million/Ton			FRP Hatcheries In Nos	
		Existing	Additional Requirement	Spawn in Millions	Fry in Millions	Carp Brood Fish in Tons	Existing	Required
1	Andhra Pradesh	87	224	42031	12609	841	0	0
2	Arunanchal Pradesh	9	4	403	121	8	19	20
3	Assam	470	0	5621	1686	112	51	0
4	Bihar	183	108	9069	2721	181	0	0
5	Chattishgarh	70	0	5900	1770	118	7	0
6	Gujurat	6	182	9883	2965	198	4	0
7	Haryana	31	40	2041	612	41	0	0
8	Himachal Praedsh	17	6	514	154	10	0	15
98	Jharkhand	118	56	4958	741	99	15	50
10	Karnataka	60	60	5671	1701	113	8	60
11	Kerala	15	20	1637	491	33	3	45
12	Madhya Pradesh	90	60	9827	2948	197	8	85
13	Maharastra	88	151	9848	2954	197	9	150
14	Manipur	11	5	817	245	16	25	30
15	Meghalaya	12	5	254	76	5	42	0
16	Mizoram	14	10	844	253	17	18	15
17	Nagaland	0	2	254	76	5	8	10
18	Odisha	155	220	16994	5098	340	71	250
19	Punjab	25	6	1144	343	23	2	0
20	Rajasthan	12	0	2527	758	51	0	0
21	Sikkim	5	0	46	14	1	4	5
22	Tamilnadu	60	0	5751	1725	115	0	0
23	Telengana	9	50	5014	1504	100	1	110
24	Tripura	60	6	1334	400	27	55	0

Sl.No	Name of the State	No of Carp Hatcheries		Requirements In Million/Ton			FRP Hatcheries In Nos	
		Existing	Additional Requirement	Spawn in Millions	Fry in Millions	Carp Brood Fish in Tons	Existing	Required
25	Uttar Pradesh	268	90	14454	4336	289	12	85
26	Uttarakhand	10	0	485	146	10	6	20
27	West Bengal	520	0	24856	7457	497	18	0
		2405	1305	182177	53904	3648	386	950

Source: Swain, S.K et al. 2021.

Table 17.11: Species-wise total seed requirement for different farming systems

Species	Fingerling requirement (in million)							Fry	Spawn	Brood Fish Requirement in tons
	Pond	Reservoir	Cage	Pen	RAS	Bio-floc	Total	In Millions		
Carp	16916	10199	0	245	--	--	27360	54719	182397	3648
Pangas	2185	--	313	--	630	1582	4710	14129	9419	471
Tilapia	--	--	--	--	66	554	619	--	688	172

Source: Swain, S.K et al. 2021.,

Table 17.12: Estimated installation of Sea cages and Requirement Fish seed thereof.

Number of cages (6m dia.) to be installed in coastal waters	Total: 20,000 cages of 6.0m dia. (Mainland 13,000; A&N Islands 6000 Lakshadweep 1000) Total Volume (@ 140m^3/cage) = 2.8 Million m^3
Cage break-up	8,000 cages of cobia (Production 2.5 tonnes/cage) 12,000 cages of other species (production 2.5 tonnes/cage)
Projected Production @ 2.5 tonnes /cage/crop	~ 50,000 tonnes at the end of 10th year (Present production through cage farming ~ 10,000 tonnes/year)
Seed requirement	8.0 million cobia seeds; 51.0 million seeds of other species (stockable size)
Number of hatcheries	8 cobia hatcheries (1.0 million capacity/year) + 25 hatcheries for other species (2.0 million capacity/year)

Source: Dr R Jeykumar, CMFRI- Personal communication

Seed Requirements for mariculture including cages has been calculated and projected in Table-12. In order to achieve the targeted installation of cages we may require 80 Crores of fingerlings of Cobia, Seabass, Pomanao, groupers and other miscellaneous fishes and shell fishes as 2 Crores of crablets/PL etc.

Keeping in view of the demand and supply of the Marine Fish seed for undertaking mariculture it is high time that a **National Marine Fin Fish Brood bank** (NMFBB) is very much essential with technical support of listed institutes and expansion through network system to ensure seed production and supply of commercially important marine fish species such as Cobia, Seab b ass, Silver & Indian Pompano, Red snapper, White snapper, Emperor etc.

'Fish Feed Technology and production

Many ICAR institutions have developed feed for different species under culture and based on their nutritional requirement. ICAR-CIFA has developed carp brooder, carp starter, magur larval diet and magur grow-out feed, pangas feed etc. Generally, it is observed that the profit by using handmade or farm made feed in grow out culture is very minimum and taste of the fish is also not up to satisfaction of customers. Therefore, application of commercial formulated feed for inland fish farming is very requisite. In order to have a sustainable fish production using of formulated commercial feed is very essential which ensures better growth, improved FCR, higher palatability to fish, disease resistance and sustained fish production.

Similarly, ICAR-CIFA has also developed CIFABROOD™ which is an exclusive carp brood stock diet, adequately rich in essential nutrients. It advances gonadal growth and maturation, facilitates early spawning and significantly increases spawning response. The feed is also suitable for multiple/ repeated breeding, off season gonadal growth and post spawning recovery. The rate of feeding is 2-3 % of total body weight during egg development phase. It is very essential that the hatchery owners develop their own formulated nutritionally balanced diet or buy the commercially brood feed from the market which will give them quality and higher seed production.

There are 38 feed plants in India that mainly manufactures shrimp feeds, with a total capacity of 3.5 million MT. In 2019, the volume of shrimp feed deals was assessed at 1.3 million MT. Andhra Pradesh represents the largest market for shrimp feed and is expected to account for 53.90% of the total shrimp feed market by 2022-23 followed by West Bengal (13.20%), Gujarat (11.0%), Tamil Nadu & Puducherry (10.50%), Odisha (7.40%) and Maharashtra (2.00%). (Dinesh & Varatharajan, 2022).

Health Management

Diseases pose a major threat to aquaculture, particularly in intensive systems and poorly managed aquaculture ponds. Further, the poor knowledge of the fish farmers and availability of quality seed, feed and medication near to the farms add to the owe of the farmers. It has been estimated that diseases account towards economic loss of around 10% of the production cost. Diseases viz., aeromoniasis, edwardsiellosis, columbaries, argulosis, dactylogyrosis, ichthyophthiriasis, tilapia lake virus infection in food fishes, and koi herpes virus and carp edema virus in koi, cyprinid herpes virus in goldfish, etc. pose serious concern in aquaculture. Apart from that shrimp aquaculture has been severely affected by the dreadful viruses. More than 20 viruses have been found deadly in shrimp itself. However, research are still on how to overcome the problem faced due to diseases. Fish Pathologists are working hard to overcome the issues faced due to diseases. National Surveillance Programme on Aquatic Animal Diseases (NSPAAD) program of Govt of India is underway and it is a platform through which farmer and scientist can detect many deadly diseases in the sector. However, the best formula for farmers is "**Prevention is better than cure**".

System diversification in aquaculture

Re-circulatory aquaculture system (RAS) is a new and unique way to culture fish in a controlled environment to maximize the fish production. In this system, through the limited space requirement, improving water use efficiency, higher stocking density, reduced manpower, ease to feed, grade and harvest the fish and the systems filter and clean the water for reuse with zero water exchange fish can be cultured. RAS are beginning to become a popular and well-accepted technique for the culture and sustainable management of aquatic animals. Though the system is very risky due to its intensification but at the same time the system has a wider scope in the days to come as the system requires less water rather it recirculates the water without wastage.

Bio-floc technology

Bio-floc is defined as macro aggregates composed of diatoms, macro algae, faecal pellets, exoskeleton, remains of dead organisms, bacteria and invertebrates. The bio-floc community also includes floc grazers such as some zooplanktons. The particles in flocs are agglutinated by bacterial enzyme-rich polysaccharide slime and held together by electrostatic attraction. Typical flocs are fine, highly porous (up to more than 99% porosity), irregular by shape, have a broad particle size distribution, easily compressible, permeable to fluid. In this system, a co-culture of heterotrophic bacteria and algae is grown in flocs under controlled conditions within the culture pond. If carbon and nitrogen are well balanced in the solution, ammonium in addition to organic nitrogenous waste will be converted into bacterial biomass. Nitrogen control in bio-floc pond is induced by adding carbohydrates into the system, and subsequent uptake of nitrogen by heterotrophic bacteria which resulted in the synthesis of microbial proteins through assimilation thereby removing ammonium from the water. This promotes nitrogen uptake through immobilization of ammonium by heterotrophic bacterial growth, which decreases the ammonium concentration more rapidly than nitrification as the growth rate and microbial biomass yield per unit substrate of heterotrophs are a factor 10 higher than that of nitrifying bacteria. Further, during the process of ammonia conversion to microbial proteins, less dissolve oxygen is required when compared to nitrification. Use of this technology aspires for zero or minimal water exchange, target to achieve maximal biosecurity in the pond and attempts to minimize external environmental effects. Some of the major benefits the technology offers to the culture practices are improvement of water quality, serve as nutrient for fish, act as antimicrobial agent and increase the fish production. Bio-floc technology provides high productivity, Low feed-conversion ratios and a stable culture environment. Bio-floc technology can help deliver sustainable production at lower cost. (Swain and Ferosekhan, 2022).

However the technology has its own limitation in our country as our country being tropical country and have multiple fish species undertaken for culture under trails by farmers. The research institute needs to standardize the species specific Bio-floc culture practices so that the farmers or entrepreneurs don't incur losses in the culture practices.

Aquaponics system

Aquaponics is the combination of recirculating aquaculture and hydroponics in a single production system. In an aquaponic unit, water from the fish tank cycles pass through filters, plant grow beds and then back to the fish. Secondly in this system, the aquaculture effluent is diverted through plant beds and not released to the environment, while at the same time the nutrients for the plants are supplied from a sustainable, cost-effective and non-chemical source. This integration removes some of the unsustainable factors of running aquaculture and hydroponic systems independently. In this production system it combines fish farming with soil-less vegetable production in one recirculating system. Nitrifying bacteria converts fish waste (ammonia) into plant food (nitrate). The same nitrification process that happens in soil also happens in the aquaponic system. The most important part of aquaponics, is that , the bacteria, is invisible to the naked eye. The key factors for maintaining healthy bacteria are water temperature, pH, dissolved oxygen and adequate surface area on which the bacteria can grow. (Swain and Ferosekhan, 2022).). This production system is gaining popularity day by day as homemakers are now undertaking these practices on roof tops of their building.

In-Pond Raceway System for intensive fish production

In-Pond Raceway System technology was developed by Auburn University, USA and American Soybean Association, USA. In-Pond Raceway System (IPRS) is a strategy to increase fish production with minimal environmental impact. The IPRS's concept is to concentrate fish in concrete cells (10% area of the total pond) or raceways within a pond and the cells are provided with constant water circulation to maintain optimal water quality and to improve feed management. Water circulation, mixing and aeration are two critical elements to this approach because it accelerates assimilation of organic loading of the pond feeding the fish. Some of the challenges in this system is that this system faces potential for disease outbreaks due to higher densities in some species and the need for having constant back up electrical energy. In IPRS, high density stocking (5000-10000 nos per 100 m3) is possible and Improved fish production (5-10 tonnes per raceway cell) has been demonstrated. It has lot of advantages i) reduced production cost (30 to 40% less), ii) Improved survival and easy management, iii) improved feed conversion ratios (FCR: 1.2 to 1.5) and feeding efficiency, iv) more than one species can be cultured in different raceways, v) less water discharge and easy sludge removal, vi) easy harvesting procedure and reduced labour cost for farm operation and vii) sludge can be used for horticulture or floriculture. The major constraint in IPRS is high capital investment (nearly 1 crore per ha), and it requires continuous power supply and requires highly skilled technicians.

Issues and challenges in Freshwater Aquaculture

Few issues and challenges in freshwater aquaculture are highlighted below based on my experiences.

- Problems in controlling inbreeding and mix breeding of fishes in hatcheries. Lack of Quality Brood stocks of commercial species.
- Unauthorized and illegal introduction of exotic species.
- Breeding and transportation of banned species like bighead carp (*Aristichthys nobilis*), Thailand Magur (*Clarias gariepinus*) and other species.
- Weak enforcement of policies for fisheries for protection, conservation, and propagation of brood stock and juvenile species during breeding season.
- Poor technology adoption and adaptation by the rural unprivileged community due to social issues and economic concerns.
- Lack of Fish seed policy in the country.
- Non availability of regulatory body for FW Inland aquaculture
- Lack of regulated market impacts profitabilty to farmers in commercial fish culture.

Issues and challenges in Brackish water aquaculture

- Enhancement in production for meeting demand
- Diversification - Satisfying every one's taste
- Sustainable production and resources conservation
- Climate change and anthropogenic vagaries
- Attract Farmers: Improved price realization
- Increase employment for inclusive development and empowerment of fish farmers

- Non availability or Less availability of disease-free Quality Seed
- Less seed (PL) survival.
- Slow growth- High FCR.
- Threat of viral diseases outbreaks
- Non-Availability of SPR/SPT stock in India.
- • Challenges includes fluctuating market prices, rising input costs, etc

Issues and challenges in Marine Fisheries in India

- Poverty; the right to food.
- Overfishing or over exploitation
- Ecosystem degradation.
- Lack of sound and suitable deep sea fishing policy
- Unorganized marketing system
- Strengthening of time series data on the deep sea the resources
- Higher Capital Investment and Recurring cost
- Non availability of skilled manpower
- Intrinsic life history traits of deep-sea fishes
- Non availability of industrial Fishing boats or factory vessels for deep sea fish exploitation
- Marine litters affecting the degradation of habitats and depletion of stocks

Government Initiative

Through concerted and collaborative efforts by the government and private sectors, India's fisheries and aquaculture sectors have made remarkable advancements towards modernization and sustainable economic growth over the past few years. The Government of India, recognized the substantial potential of this industry and recently unveiled the Pradhan Mantri Matsya Sampada Yojana (PMMSY), a new flagship programme with a significant investment of Rs. 20,050 crore (US$ 2.46 billion) for the growth and expansion of fisheries sector in the country.Apart from that Rs.7,520 crores for Fisheries Infrastructure Development Fund (FIDF) also was another boost to the sector.

The PMMSY scheme aims to address significant productivity and production gaps in the fisheries sector, infuse innovation and cutting-edge technology, enhance post-harvest infrastructure and management, modernize and strengthen the value chain and traceability and establish a framework for effective fisheries management and the welfare of fishermen. The scheme highlights the importance of leveraging fisheries potential in a sustainable, ethical, inclusive and equitable way.

The PMMSY lays out ambitious goals for FY 2025, including increasing fish production by 70 lakh MT, increasing aquaculture productivity from the current national average of 3 tonnes per hactre to 5 tonnes per hactre, doubling exports from Rs. 46,589 crore (US$ 5.72 billion) to Rs. 100,000 crore (US$12.28 billion), creating more than 55 lakh job opportunities, and doubling fishers' and fish farmers' incomes to consolidate the fisheries' sectoral gains and boost economic growth.

Way forward to increase the aquaculture production in the country

The Indian fisheries and aquaculture sectors have become the fastest growing agriculture allied sectors in the country. The share of inland fisheries and aquaculture in overall fish production has gradually increased over the last several years. India has also become top producer and exporter of shrimp as a results of using contemporary technologies and several Govt initiatives New areas, such as salty and brackish waterways, reservoirs, and flood plains, are being used for profitable aquacultures. wastelands are being transformed into wealthy lands.

The government has implemented several supportive measures to increase the potential of these sectors. It is developing fishing infrastructure, landing areas, river ranching, and fish seeding on major rivers, including the Ganga, the Brahmaputra, and others. In the Himalayan and North Eastern states, cold water fisheries, trout brooding facilities and hatcheries are being developed.

The Government aims to increase fish production from 140 lakh tonnes to 220 lakh tonnes by FY 2025, which can be achieved through marine fisheries, inland fisheries, and aquaculture. The induction of deep-sea vessels, modern and effective fishing methods, a network of seamless cold chains, cold storage facilities, ice plants, fish processing units, fish product units and many more have been envisioned. Given the positive developments, it is reasonable to believe that India's fisheries and aquaculture sector have a bright future.

However, few lines are highlighted for sustainable aquaculture in India : a) Seed production & associated infrastructure (similar to private sector multi species hub facility at Odisha) should be developed or the existing facility needs to be upgraded in strategic locations of the country on a priority basis to create a robust fish seed value chain that reaches all the potential aquaculture areas. b) Fish seed village concept similar to Pakhanjure of Chhattisgarh may be adopted widely meeting the increase demand of the sector. c) Mean productivity level of the existing aquaculture ponds remains as low as 3.0 t/ha should be easily increased to 5.0 t/ ha through adoption of scientific freshwater aquaculture practices. d) Intensive aquaculture technologies with genetically improved fish species to be adopted in the different systems for higher fish production. e) To promote the widespread adoption of the technologies, systematic demonstration programmes to be taken up involving all the relevant stakeholders. f) Many panchayat water bodies and revenue ponds are underutilized across the country, which should be brought under scientific aquaculture in a phased manner involving the women SHGs and similarly smaller reservoirs are also to be utilized judiciously. Proper convergence to be ensured with the other ministry/departments to forge a synergistic partnership to utilize the waterbodies for fish production. g) Capacity building should be given top priority; a greater number of farmers should be trained on specific technology modules on scientific aquaculture. Sufficient training infrastructure similar to Aqua field schools should be developed in all the states h) Many states lack proper infrastructure facilities like fish & shellfish hatcheries, soil - water quality testing labs & disease diagnostic labs. All the potential states should be advised to establish the labs at least in every district having major aquaculture operation hubs. i) Domestic fish consumption should be promoted aggressively as it would ensure the sustainable production of fish from freshwater aquaculture sector j) Creating an enabling environment for attracting private investments in the sector through pro-active, innovative public-private partnership business models with viable technological solutions involving all the relevant stakeholders & developmental agencies k) Mission mode schemes to be implemented in high potential species like scampi, ornamental fisheries, and freshwater pearl farming to promote export-oriented entrepreneurship development. l) Institutional reforms for effective governance structure in line with the dynamics of the sector should be given priority m) Robust impact assessment framework to be inbuilt in the implementation of the scheme along with a scope for need based research studies to increase the operational efficiency of the scheme implementation process n) Exploitation of untapped deep-sea fishery resources through induction of modernized deep-sea vessels. o) Ecosystem-based fisheries management needs to be adopted as it is a holistic approach that recognizes all the interactions within an ecosystem rather than considering a single species or issue in isolation p) Conserving the over exploited or depleted coastal resources and recruiting through sea-ranching on priority basis p) Implementation of uniform marine fisheries policy including deep sea fisheries policy in the country is the need of the hour.

Conclusion

With a view to bring blue revolution in India, Govt of India has emphasized several programmes to enhance fish production through horizontal expansion and vertical increase in terms of resource renovation and management, system & species diversification, breed improvement and technology interventions. However, brood stock upgradation, national seed dissemination programmes, round the year production of fish seeds, standardisation of seed production of few more commercial species, farm made feeds prepared from locally available ingredients, disease surveillance and remedy to prevent and control diseases, impact assessment of technology etc. are the major aspects need to be addressed. Capacity building of human resources including scientists, farmers and other stakeholders is always a priority to us in order to be globally competitive and ensure food and nutritional security of the country. By 2050, India has to produce about 20 MT of fish/year and the 80% of the total production must come from the aquaculture Therefore, a way forward should be to evolve genetically improved varieties of fish and shellfish through an integrated cost- effective marker assisted selection and genomic selection programs at national level for prioritized species. The sustainable development in aquaculture is always linked with policy support and proper strategies for technology transfer and adoption for effective resource utilization, market linkage, post-harvest product processing, etc. We hope to succeed in all fronts with the given strategies and targets. A robust Fisheries Policy in all the three sectors is the need of the hour and it will not only increase the production but also will have control over the ecosystem and market including traceability for food safety as human health is the first priority.

References

Anon, (2022). Hand book of Fisheries Statistics-2022, Dept of Fisheries, MOFAHD, Govt of India. pp 216.

Dinesh, R. & Varatharajan, Durai. (2022). An Outlook on Indian Shrimp Feed Industry and Manufacturers. 10.13140/RG.2.2.19253.76003.

FRAEED, CMFRI, (2023). Marine Fish Landings in India-2022. Technical Report, CMFRI Booklet Series No. 31/2023. ICAR-Central Marine Fisheries Research Institute, Kochi

Swain S.K. and Ferosekhan. S. (2022) Present Status and Future Scope of Freshwater Aquaculture Sector in India. pp.84-99 In Das, B. K. et al (2022). Souvenir of 1st Indian Fisheries Outlook, ICAR-CIFRI, Kolkata, pp 251

Swain, S.K., Pillai, B R., Das, P. C., Sivaraman I., Chakrabarti, P., Sahoo, P. Sundaray, J.K., Mahapatra, K.D., De, H., K., Mohanta, K.N., Hemaprasanth., Mohapatra, B.C., Routaray, P., Giri B.S., and Barik, N.K. (2021) Actionable Strategies for Increasing the Freshwater Aquaculture Production in India. ICAR-Central Institute of Freshwater Aquaculture, Bhubaneswar. pp.142.

My Journey with Natural Farming

Subhasisa Bal

I always feel I am very lucky and got a chance to live in this beautiful God gifted planet by blessings of God and whatever God fixed for me I am trying my best to do so that I can fulfil my journey and reach to my destine and finally will back to the same place where I came. My father was a Botany Lecturer, he retired from his job in the year 2009 and my Mother is a house wife. I have one brother and two sisters. All are married now and me my misses, my son (Calyx) staying at my native village with my parents after I came back from Indian Agricultural research Institute and my brother and two sisters are staying at Bhubaneswar. I have a very small family staying happily at my village. I have completed my schooling at my village school which is just 1 km away from my home. My village Garh Guatira is belongs to Nimapara block of Puri district of Odissha and it is just 50 km away from Bhubaneswar. My village is very small which is surrounded by agriculture field specially 90% area is under rice cultivation. So I spent my childhood with my natural village climate and maximum time I spent in the village surroundings with its natural activities. I got first class in Botany (Hons) with distinction from the Nimapara college and it was one of the best moments for me, at least one step I moved towards my destination. To get this happiness many nights did not sleep may be I was in a small college but it was my way to destination and from there I got the dream to be a scientist in agriculture but I was a basis Botany student and to understand the agriculture it was a difficult journey. From these days I got dream every day how I will get chance to do research in agriculture filed. But the dream was not possible as I was a Botany background student. When my teacher Dr.Rabindra Nath Padhi took a class on economic botany, especially when on rice that time he used the term ICAR, IARI, IRRI like that but that words fixed in my mind from that days and I dreamed it every day where is IARI ? where is IRRI ? how the scientist working there ? who is the scientist ? how they get job there ? these thing fixed in my mind and every day I think about it, that things made me mad to join IARI and IRRI not for a job but to learn organic farming, natural farming, regenerative farming.

Then to reach the destination I have completed my M.Sc and Ph.D Botany from PG Department of Botany, Utkal University, Bhubaneswar. During my Ph.D period I have suffered from brain malaria and went for coma for three month, I thought I cannot save my live, everyone feel I will die but by the blessing of God I recovered from the fever and again worked and completed my research work for my Ph.D degree. After my Ph.D degree I joined as Post Doctorate fellow at different institutes like Regional Research Laboratory (RRL) Bhubaneswar, Regional Plant Resource Centre (RPRC) Bhubaneswar, National Research Centre on Plant Biotechnology (NRCPB), Indian Agricultural Research Institute, New Delhi, Directorate of Rice Research (DRR) Hyderabad, Directorate of Oil Seed Research (DOR) Hyderabad, Consultant Scientist Hi Media Lab. Mumbai, National Consultant for Regenerative Farming (IRRI) & Founder: LANDRACE since 2015.

From my childhood it was in my mind and it was a dream, after my education I will not do job I will run a small agriculture school for kids or I will do a business related to rice or I will start something my own on agriculture so that I can provide a small employment to those who need it for their family that was in my mind from that days, so slowly when I reached higher education I got more idea to implement it and presently I am running my own organisation (Landrace) a village based small organic rice research organisation where I am working as a free bird, sometimes I have to work as a worker, sometimes I have to work as a marketing person, sometimes I have to work as a Scientist, sometimes I have to work as an owner of the organisation. My aim to save the lives as life is precious and save the planet for our kids and other lives of the planet. The journey I decided really it a challengeable job for me still sometimes I feel very happy as it is my dream but 99 %. I struggled to save my journey and to save my life, even sometimes I feel I should quit it, I feel I can't do it or I am not a right person to do this work. I do not know where I should start writing about my dream, too difficult. My father, my family members, my close ones, my friends even all those who know me very well they thought after my education I should do a job inside the AC room but I decided it from the beginning after my study complete I will do organic agriculture or natural agriculture or regenerative agriculture and will teach farmers how by adding different agrochemicals we are destroying our nature which will not suitable home for our kids and other life forms. only to save this beautiful earth from different dangerous chemicals, I leave AC laboratories and my job but I don't know my passion one day will keep me away from everyone, as for them I am not a educated person I am just a worker, I am just a mad, mentally disturbed person, even some times those don't know the value of agriculture they misbehave me very strongly which was so painful to tolerate, but my dream is to give a handful of chemical free food on everyone's plate and a dream to save the beautiful planet earth from dangerous chemicals.

My father has near about 10 acres of agriculture land and before me he was involved in rice cultivation through chemical farming when he gets free time from his lectureship job. Exactly 2015 I have approached my father to do organic rice production in that land he denied to do that he told me to go back to Delhi and whatever want I do there only, do not came back to village, as everyone know that I am doing job at Delhi, that time I already have left my job, so that time I was hopeless after listening this things from him, so I did not understand what I should do. Some days I move away from my home again to Delhi, as for me my parents feel a headache, for them I was a unwanted person or even for all I was an unwanted person, that time was very crucial for me how I can manage all the things and my life, without land it was not possible.

Time passes away I was able to convenience my father, he gave me just one acre land, from that one acre I have standardised organic rice production and then expand to 10 acres in my own village. Apart from that presently I also expanded my aromatic rice cultivation near about 90 acres with more than 150 farmers involvement at different part of Odisha and India. Presently I am cultivating near about 70 different shot grain aromatic nutritional rice like Kalajeera, new kalajeera, chinikamini, new chinikamini, new dhusara, gobindabhog, krishnabhog, radhabhog, tulaipanji, sugandha, harinarayana, kalikhasa, dubaraj, jayphula, maliphula, kalanamak, sorisaphula, gitanjali, etc. I am also connecting farmers directly with the consumers so that both will be benefited, same time a farmer will not get any problem to sale their products and at the same time a consumer will get a better product according to his choice for his family.

I face many problems during my work due to climate changes, 2018-2019 was a bad period for me, I spent all my savings in agriculture that year but due to cyclone 'Funy' all my crops gone, and all my savings finished, and all my important rice landraces gone. I remember once I had zero balance in my account but I never quit. That year still I remember, so horrible situation for me to manage my work and to manage my day-to-day life. Even I was not able to purchase a pair of sleepers for me, I used old sleepers which condition was very bad but I don't have option that time but I managed everything by closing my eyes and keeping a big stone on my heart, I am unable to describe that situations so painful for me even if I am remembering those days tears are coming to my eyes. That time my close ones laughed at me and they thought I am a mad. With time every bad time pass away and I am able to make a bridge between producers and consumers and promotion of residue free rice and rice based value added products for better health and better nutrition to control the malnutrition among kids, all age groups and to develop a better immune system to fight against diseases and for better gut health and to save the planet biodiversity. I finally feel to do something it need lots of patience, lots of sacrifice, lots of dedication, lots of faith, lots of self-confidence, lots of dream then the way to destination will open. Now I am so happy that people connected with me from different part of the country and scientists from other countries. I presented my work at SAARC platform where scientists from SAARC countries appreciated my work and they want to implement my research in their countries to save the mother planet. I am herewith sharing some of my farm snaps and how people connected with me from different part of the country and outside of the countries to learn my 5G Farming technology.

Jai Jawan Jai Kisan jai Vigyan

My Smart Aroma Nursery

MY Village Field Research Station

5G Farming Aroma Rice

Aroma rice & BGA Conservation Site

My Family 5G Farm

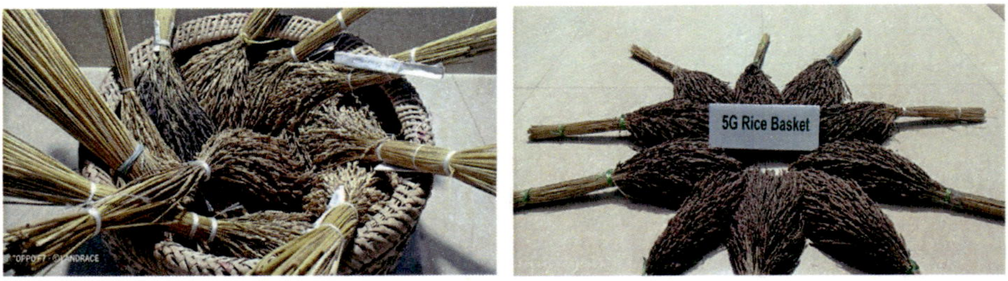

My 5G Rice Basket

5G Rice

My Kid Calyx Involved with me.

My Kid Calyx Introducing to the rice world

Finally, I bring 5G Rice for my kid

People connected with my 5G Farm Rice Aroma Mission (RAM)

Dr. Subhasisa Bal
Founder: LANDRACE
At-Garh Guatira, Po-Arisandha, Dist-Puri, Odisha, India,
M-9439471194
Email: landracein@gmail.com

My Entrepreneurial Journey with MARSCO

Mahidhar Bolem

The study of fish in the sea may be the most necessary of all our oceanographic researches because we shall increasingly be made to turn to the sea as a vast food producer by the increase in the population of the world.

— T.F. Gaskell

Aquaculture isn't just farming fish; it's indeed farming the future. I've read countless studies on how aquaculture is the future of human race.

"The enormity of the situation can be further gauged from the fact that more food has to be produced in the next 35 years than what was produced in the last 8,000 years," says Dr Vijay Gupta Modadugu.

But it wasn't those statistics that propelled me from a banker's office to the muddy banks of a shrimp pond. It was witnessing the struggles of individual farmers, their dreams as fragile as the fish they raised. Aquaculture is about more than food, it's about those dreams, about building a better future for them and generations to come. My journey with Marsco is my way of investing in that future.

Echoes of the Past

Before embarking on the narrative of 'My Entrepreneurial Journey with MARSCO', it is imperative to revisit my humble beginnings and the path that led me here. Born nearly 50 years ago in Vijayawada, Andhra Pradesh, I hail from a community that faced both economic and social disadvantages. Despite these challenges, I was determined to carve out a different future for myself.

My academic journey began with a Bachelor of Arts in Mathematics, Economics & Politics from Andhra University, a stepping stone towards my ultimate goal - a Master's in Business Administration (MBA). This pursuit was not merely a career choice, but a tryst with destiny. My academic endeavors didn't stop there. I went on to earn a Doctorate in Management, specializing in Pharmaceutical Marketing, and also became a certified Business Analyst.

This journey, however, was not just about acquiring degrees and professional qualifications. It was during these formative years that I encountered a profound quote by Mahatma Gandhi on the 'Seven Deadly Sins': wealth without work, pleasure without conscience, knowledge without character, commerce without morality, science without humanity, worship without sacrifice, and politics without principles.

Regarded as one of Gandhi's significant contributions to ethical philosophy, the concept of the 'Seven Sins' intrigued me early on, not merely for its cautionary message, but for its awakening. The dichotomous implications of the quote piqued my curiosity. I observed that all of the 'seven sins' are tied to social and political circumstances. Perhaps, the remedy for each of these "deadly sins" lies in an explicit external standard or something rooted in natural principles and laws, rather than societal values.

Given my interest and career aspirations in 'commerce', Gandhi's quote prompted me to ponder - are business and ethics fundamentally at odds? Are they really the binary oppositions? This is the beginning of my exploration into this question.

I specialized in the strategic planning and execution of business goals, fostering team building and leadership, driving development and negotiations, managing debt and risk, cultivating relationships with high-profile clients, and re-engineering business processes for optimal efficiency.

In the late 90s, I started my career journey as a Team Leader at Eureka Forbes Ltd. After a couple of years, I moved into Sales & Marketing at GlaxoSmithKline Pharmaceuticals Ltd. As the new millennium began, I took on a role as a Zonal Sales Manager for Alembic Limited's North & West Zone, which I held until the mid-2000s.

Painful Precursor

A major shift in my career trajectory occurred in the mid-2000s when I stepped into the role of Regional Debt Manager for South 1 DSMG Agri at ICICI Bank Ltd. My primary responsibility was to curtail the company's Gross credit loss. This involved handling the debt service of various rural products such as Farm Equipment, Farmer Finance (Personal, Poultry, Aquaculture, Horticulture, Cattle etc.), Rural Housing, Loan against Property, Agri Credit Line, and Small Agri Loans. This role required me to interact with farmers from various sectors - agriculture, horticulture, aquaculture, and more. It was a role filled with unique challenges, but also one that offered the opportunity to overcome them.

One day, I found myself face-to-face with Emmanuel Rebba, an aqua farmer and loan defaulter. While such issues were typically resolved at the field staff level, my reflections on the intersection of business and ethics, sparked by Gandhi's teachings, often led me to personally visit those grappling with debt.

Rebba was a weathered man, his hands bearing the marks of countless hours under the sun and in the saltwater. Upon my arrival for the loan assessment, I was met with a somber scene. His ponds, once teeming with fish, were now silent and still, a graveyard of his dreams. Empty feed bags lay scattered around, stark reminders of the failed harvest. Rebba stood by the edge, his despair mirrored in the faces of his family watching from a distance. The desperation was palpable; this was not just a financial crisis, but a man wrestling with the collapse of his dreams and livelihood. He couldn't afford medicine, he couldn't afford to restock, and failure was not an option, especially with a daughter on the brink of college. Rebba had weathered disease cycles before, but this time, he was up against the cold, hard numbers on the bank's spreadsheet.

Caught between my duty to the bank and the crushing weight of Rebba's predicament, I felt a deep conflict. Rebba's plea for an impossible extension, delivered in a rustic tone that bordered on argumentative, only amplified this feeling. Rebba was not an isolated case. The faces of other defaulters - Samudraveni of Nagayalanka, Mastan Rao of Repalle, Malakondaiah of Bapatla - are etched in my mind, each one a stark reminder of the human element in the world of commerce.

In my encounters with aqua farmers like Rebba, I've seen the harsh realities they face. Their struggles aren't born out of laziness, but rather a lack of access to quality resources and knowledge. Poor quality inputs often lead to failed outputs, while disease outbreaks can wipe out entire ponds. Manipulative middlemen exploit their desperation, further deepening their debt. Those Aqua-farmers are not just battling financial crises, but also fighting to keep their dreams and livelihoods alive. Indeed, each face served as a stark reminder of the human element in deprivation, highlighting the urgent need for empowerment. It was during this time that the seeds of my motto - 'Empowering Aqua-Farmers' - were deeply sown in my mind.

Thus, my entrepreneurial journey with MARSCO found its genesis in these experiences. This journey is not just about business, but about making a difference in the lives of those who are often overlooked by the commercial world. It's about recognizing the human element in commerce and striving to uplift those in need. This is the ethos that guides MARSCO and shapes my mission.

Sustainability in Aquaculture!

There came a moment when I realized the importance of understanding the social, economic, and cultural aspects of the aqua-farming communities. This realization led me to immerse myself in their world, studying the natural and human resources at their disposal, the technology they required, the market demands they faced, and the issues and constraints they encountered in technology development.

My formal education and certifications in Marketing and Business Administration were evident physically. Whereas, the grassroots-level research that truly enriched my understanding, may not have the certified endorsement. My aqua- research may not have come with a certificate, but it was driven by a singular agenda - to understand the needs and constraints of those farming communities.

During the initial days of my study, I was jolted by a major criticism of aquaculture's ecological impact. This criticism served as a stark reminder of the complex interplay between commerce, ethics, and the environment.

Aquaculture has often been criticized for its environmental impact, with concerns that its growth could lead to more harm than good. While it's true that the early stages of aquaculture in the 1980s did cause some environmental damage, particularly in the case of brackishwater aquaculture where mangroves were destroyed, the industry has come a long way since then.

Take, for example, the case of brackishwater aquaculture in Coastal Andhra Pradesh during the 1980s and 1990s. Shrimp culture began to scale up commercially during that time, leading to a rapid expansion of shrimp farming. However, this growth came at a cost. The destruction of mangroves and other environmental issues were prevalent during that period.

But as the industry evolved, so did its understanding and respect for the environment. The realization of these problems led to changes in practices and regulations, significantly reducing the environmental impact of aquaculture. Later, by the time I started my concentration on aquaculture, the focus is on developing it in a more sustainable and environmentally friendly manner.

So, while the shadows of the past still linger, the industry has made significant strides in minimizing its environmental footprint. The lessons learned from the past have paved the way for a more sustainable future in aquaculture. The aqua-forming community understood that its survival depends not just on economic profitability, but also on environmental sustainability. And that was a step in the right direction.

Between Ideals & Reality

Although my ideals propelled me towards my destiny, reality, with its tentacles of own agenda, dragged me to continue in my existing profession. Consequently, my aspiration to work closely with the aqua-farming community was postponed for practical reasons.

I continued my tenure with ICICI as the Regional Manager for the South, SARFESI Cell SME & CCMG, and the Mortgage Assets Disposal Head for AP Operations. Post-2013, until the inception of my own firm, I served as a Senior Branch Manager at S My role involved managing the bank's interests when businesses in the South were unable to repay their loans. This entailed strategizing on how to recover the money – sometimes by seizing what they had pledged as collateral, and at other times, by restructuring the entire loan. I also managed large loan accounts, ensuring smooth operations and preventing losses for the bank. This role demanded extensive analysis, paperwork, and people management – dealing with stressed borrowers and ensuring our teams adhered to the rules.

In the midst of a seemingly secure life, where the comfort of a stable income from ICICI and the settled rhythm of daily existence enveloped me, I found myself at a crossroads. It was a life many would aspire to, marked by a respectable salary of nearly Rs. 2.5 lakh a month, the joy of watching my children grow, schooling in top notch institutes, and the predictability of routine. Yet, beneath the surface of this tranquility, there lurked a restlessness, a yearning for something more profound than the accolades of a conventional career path.

A Journey from Comfort to Purpose

It was during this period of introspection that the words of Pablo Neruda resonated with me like a clarion call to action. His poem, 'You start dying slowly', a beacon in the night, spoke of the slow death incurred by not pursuing one's dreams, by remaining tethered to the safe harbour of familiarity and security. Neruda's stanza echoed the sentiments buried deep within my heart—

"You start dying slowly

If you do not change your life
when you are not satisfied with your job, or with your love,

If you do not risk what is safe for the uncertain,

If you do not go after a dream,

If you do not allow yourself,
At least once in your lifetime,
To run away from sensible advice…"

This poignant reflection served as a catalyst for a profound internal awakening. It illuminated the stark reality that life's essence is not merely to exist within the confines of comfort and predictability but to embrace the uncertainty of pursuing one's passion. The realization that true fulfillment stems from following one's dreams, even at the expense of leaving behind the security of a well-trodden path, became the cornerstone of my decision-making process.

Thus, I found myself at a pivotal juncture, where the allure of a secure lifestyle clashed with the compelling desire to advocate for the welfare of aqua farmers. The plight of these hardworking individuals, their struggles and dreams, resonated with me on a profound level, challenging me to step beyond the boundaries of my own life and to commit to a cause greater than myself.

As the Senior Branch Manager of ICICI in the Siddiamber Bazar branch in Hyderabad, I found myself in an atypical banking environment. The area was a bustling business hub, and the clientele had unique requirements reflective of their commercial interests. In my role, I was tasked with increasing branch profitability, which necessitated maintaining a strong rapport with our customers, who were predominantly businessmen.

In my previous roles, my interactions were largely with the economically disadvantaged, farming communities, and lower-income groups. These interactions, in some way or other, kept the flame of my enduring dream alive - the dream of empowering the marginalized, particularly the farming community. However, the shift in my job profile, from dealing with the underprivileged to interacting with businessmen, led to a period of introspection. It made me question the path I was on and how it aligned with my deeper aspirations.

The Turning Point

At this juncture, I seriously contemplated leaving my 'secured' job to establish my own firm. My father, during this pivotal time, was a pillar of support, encouraging me to pursue my dream. Equally instrumental was my spouse, who propelled me towards these new and adventurous ventures. However, many of my relatives and close ones expressed reservations about my decision, attempting to dissuade me from this path.

In defiance of the pragmatic advice from well-wishers, I chose to venture into the unknown, to forsake the Sisyphean cycle of security for the unpredictable journey of entrepreneurship. This decision was not taken lightly; it was a deliberate choice to chase a dream that promised no guarantees, save for the promise of pursuing a passion that gave my life a deeper sense of purpose.

The leap from the safety of a well-established career to the tumultuous waters of entrepreneurial endeavour was a testament to my belief in the transformative power of dreams. It underscored my commitment to not just exist but to live fully, driven by the conviction that the essence of life is found in the pursuit of one's passions, regardless of the inherent risks.

An important anecdote comes to mind, one that helped me navigate a Shakespearean 'to be or not to be' dilemma. I was approached by old friends who had been deceived by a fraudulent company overseas. I assisted them in successfully recovering their payment. During this time, these friends began to insist that I join their firm, RR Health Care. In reality, it was an invitation to 'rejoin', as I had initially been part of the team that established the firm, which was dedicated to the health of animals and plants.

Since my childhood, I have harbored a deep compassion for animals and plants. At the onset of my career in 1996 as a marketing salesperson at Eureka Forbes, and before my tenure as the Zonal Sales Manager at Alembic Pharmaceuticals, I worked with Glaxo Agrivet Farm Care (Virbac), a major player in the animal husbandry sector. My five-year stint in Veterinary Sales with the firm rekindled my passion for animals.

When the opportunity arose to start a firm with two like-minded friends, I was presented with an attractive opportunity at ICICI. Consequently, I joined ICICI while my friends continued with their venture. Reflecting on this journey, they invited me to leave ICICI and join them.

That chapter of my life is a narrative of courage and conviction, a journey fueled by the desire to make a difference in the lives of those who have been marginalized. It is a testament to the belief that true satisfaction and fulfillment come not from the material comforts of life, but from the relentless pursuit of one's dreams. It was during this period that the verses of Pablo Neruda served as a catalyst, propelling me towards a new path. I finally, in 2014, ventured to join RR Health Care, which has Animal and Poultry units. Given my expertise in aquaculture, I added 'RR Aqua Health Care' to the firm's portfolio. This marked a significant milestone in my journey, aligning my professional path with my passion and commitment to the welfare of aqua farmers.

Due to professional differences, my two partners decided to part ways. This development presented an opportunity for me to take a significant leap forward. In 2018, I established my own firm - MARSCO. This marked a new chapter in my journey, allowing me to fully realize my vision and commitment to the welfare of aqua farmers.

Influence & Inspiration

Before I delve into my journey with Marsco, I must express my deep gratitude for the influence of the eminent biologist and fisheries scientist, Dr. Modadugu Vijay Gupta. I first learned about him in the mid-2000s when he was awarded the prestigious World Food Prize, an international recognition often equated to the Nobel Prize. His work, which significantly improved the quality, quantity, and availability of food worldwide, left a profound impact on me.

Dr. Gupta's pioneering efforts in creating an aquaculture system tailored to the needs of poor, rural populations in Asia, Africa, and the Pacific resonated deeply with me. His field-oriented research extended beyond our country's borders to Laos, Thailand, Vietnam, Bangladesh, and beyond. It was through Dr. Gupta that I first encountered the term 'Blue Revolution'. His advocacy for the food and nutritional security of the rural poor aligned with my own intuitive belief that more contributions were needed in this area.

While Dr. Gupta was serving as the Assistant Director General at 'WorldFish', an international fisheries research institute based in Malaysia, I attempted to visit him, but my efforts were unsuccessful. However, in later years, I had the opportunity to meet him on several occasions. I expressed my admiration for him and shared my dreams and humble efforts to turn a vague idea into concrete action.

Dr. Gupta's influence ultimately shaped my career path with Marsco, inspiring me to create innovative ways to empower the rural poor in my vicinity by developing and disseminating techniques for low-cost fish farming. His work served as a beacon, guiding my efforts to make a meaningful impact in the lives of those who are often overlooked.

My journey towards becoming a certified Master Trainer - PMT for Poultry, Dairy & Aquaculture from NIRDPR, Hyderabad, and a Trainer of Trainee from

CIFA, Bhubaneswar, was significantly influenced by Dr. Vijay Gupta. His work and philosophy also guided me as I ventured into the field of microbiology, specializing in wastewater and solid waste treatment, the aquaculture industry, septic tank treatment, and bioremediation of ponds, lagoons, and freshwater bodies with the help of microbes. Whether directly or indirectly, Dr. Gupta's influence has been a guiding force in my professional journey, shaping my approach and reinforcing my commitment to making a meaningful impact in the lives of aqua farmers. His teachings continue to inspire me as I navigate my path with Marsco, striving to empower the rural poor through innovative and sustainable aquaculture practices.

The Leap of Faith

When I decided to venture into this sector with my own business, I noticed a common trend: the same product was being used across different industries, be it cattle, poultry, or fish. Regardless of the product, we make it a point to consult with industry-specific professors and nutritionists. Only after these consultations do we proceed to develop a product. Our commitment to creating products backed by cutting-edge technology has been a significant driving force for our company.

In the aquaculture industry, one of the most significant hurdles is cost recovery, given the sector's largely unorganized nature. Initially, our identity was unknown, and we had to establish our reputation in the industry. To gain traction for our products, we had to extend credit to the farmers, which led to initial financial losses. However, we subsequently refined our payment policies and lowered our prices to make them more accessible to farmers.

I don't subscribe to the notion of competing in a numbers game or striving to be at the top. Instead, my goal is for farmers to regard Marsco as their preferred company and their first choice for their needs.

I've explored numerous concepts aimed at enhancing farm productivity, improving supply chain efficiencies, and boosting farmer profitability through more effective market linkages. This iterative process is crucial in identifying the most impactful strategies and solutions.

My ample experience in the aquaculture and animal husbandry industry, coupled with my ongoing efforts to understand the marketing models in this sector, will undoubtedly contribute to the company's growth and success. This knowledge and understanding are key to driving the company forward.

The Journey with Marsco

In my role at MARSCO, I've been collaborating with stakeholders in the Indian aquaculture value chain. Our collective efforts are focused on enhancing farm productivity, establishing better market linkages, and facilitating access to formal financing. This work is crucial in supporting the growth and sustainability of the aquaculture industry in India.

Marsco Nutraceuticals, the company certified by ISO 9001:2015, WHO-GMP, and HACCP, has established itself as an Indian-based firm specializing in the manufacturing, marketing, and technical consulting of Animal & Aquaculture feed supplements. Recognized under the #StartupIndia initiative with the identifier DIPP78628, it is also an incubatee of the RKVY-RAFTAAR programme. This recognition and affiliation underscore the company's commitment to quality and innovation in the industry.

Marsco Aqua Clinics - Aqua One Centre has been appointed as a nodal agency by the NFDB and MANAGE. Our mandate includes conducting awareness campaigns, training, capacity building, and skill development programs, both online and offline. As I forge ahead, my primary goal is to elevate Marsco Nutraceuticals to the status of the most sought-after provider of animal and aquaculture health care solutions, both locally and globally. This objective guides our efforts and shapes our strategies.

The mission of Marsco Nutraceuticals is to provide expert opinions, technology transfer, and project management services to new and existing projects nationwide. We are committed to delivering high levels of dedication and service to every project we undertake, viewing the success of each project as a direct reflection upon our skills and consummate reputation.

Our services are designed to address the diverse needs of the aquaculture industry. We provide bankable business plans for proposed and existing aquaculture ventures, emphasizing the importance of having a constantly updated 'living business plan.' We specialize in freshwater and marine fish, and shrimp farming projects, requiring a feasibility study prepared by our team for project management services. We advise qualified clients on investments in aquaculture projects and provide due diligence on high potential investment opportunities. We offer skilled engineering services in pond design, RAS, and more, supporting engineering expertise for various aquaculture projects.

We also provide expert opinions in various aquaculture-related legal cases with experienced experts. We emphasize the importance of proper site selection for aquaculture projects, highlighting the significance of a suitable site for the success of the project. We offer training for local labourers and aquaculture experts, preparing local management teams to take over operations smoothly.

Milestones & Achievements

As I continue on this journey, I am committed to maintaining these functionalities and thriving in my role. This commitment is reflected in the on-going success and excellence of Marsco Nutraceuticals in the aquaculture and animal husbandry industry.

In the relatively brief span of just six years, I've reached significant milestones with Marsco. The catalyst for this journey was undoubtedly Aquaculture, and seizing the right opportunities within this field.

1. **CAA Approved Products**: Rather than merely accumulating certifications, I've achieved something far more meaningful and impactful. Obtaining CAA approval for the first 20 products and achieving ISO certification felt like leaping over the first hurdle in the daunting world of aquaculture.
2. **Established Own Godowns**: Setting up godowns in Vijayawada, Jaipur, and Balasore wasn't just about marking locations on a map. It involved late nights securing deals and even sleeping in my car between site visits.
3. **NFDB & MANAGE Registered Aqua One Centre**: Building Marsco from concept to growth involved getting NFDB and MANAGE registration. It wasn't just a stamp of approval, but a testament to someone believing in my vision with integrity enough to open those doors.
4. **Own Transportation Facility**: Marsco became synonymous with timely delivery, thanks to our own transportation facility. Farmers can't wait weeks for supplies, and there were times early on when I was that delivery guy.
5. **Member of Course Empanel at Centurion University**: Being part of Centurion University and working with NFDB marked a shift from being an entrepreneur to someone creating lasting impact.
6. **Employee Network Spread in 8 States**: The incredible reach of Marsco's employee network across eight states is a testament to trust, delegation, learning from failures, and seeing potential in others. It's about a participatory approach and freedom of expression to employees to finalize strategies.

7. **Recognised as Startup India Company**: Being recognized as a 'Startup' by the Government of India validated my belief that it's never too late to start.
8. **Privilege Trips to the US and Singapore**: These trips were not about luxury, but about rewarding hard work and showing my team a world of possibilities they might not have imagined before.
9. **Capacity Building Trainings to Employees**: Some early business advice was 'be the boss, they work for you.' It's never worked for me. My team knows they can disagree, come up with better solutions than I could. It is the participatory approach.
10. **12 Publications in International Journals**: These 12 international publications mean more to me than many awards. It's about contributing knowledge to the field, not just making a profit.
11. **International & National Level Expo Participation**: By having an innovative product range, Marsco became an incubatee of national prestigious institutes like MANAGE, SKUAST & IVRI.
12. **"Krishi Seva Ratna" Award in 2021**: This award is a testament to the hard work and dedication that has been put into Marsco, not only by me, but also by my entire team.
13. **More than 25 Brands Registered**: More than 25 brands are registered under the Government of India Intellectual Property of Trade Mark & Patents.

These milestones reflect the journey of Marsco and its commitment to making a meaningful impact in the field of aquaculture.

Mentorship and Collaboration

My entrepreneurial journey is often portrayed as a solitary one, but the truth is that success rests upon a strong network of support. I've been privileged to learn from and collaborate with some of the brightest minds in the aquaculture sector. Leading figures like Dr. B.K. Das, Director of ICAR-CIFRI, and Dr. Bimal Prasanna Mohanty, Assistant Director General of Inland ICAR, have shared their invaluable insights, propelling Marsco's innovations.

I'm also grateful for the guidance of esteemed academics like Dr. Gadadhar Dash (WBUAFS), Dr. Manas Kumar Sinha (NFDB), Dr. A.M.K. Mohan Rao (Former JD, Vertebrate Pest Management), and those in leadership roles like Dr. Shahaji Phand (MANAGE), Dr. Saravanan Raj (MANAGE), and Dr. Abbaraju Udaya Shankar (KL University). Their expertise has sharpened

Marsco's mission and strengthened our industry connections.

The field of aquaculture is constantly evolving, which demands continuous learning. Dr. Jyothi Kachroo (SKUAST-Jammu) has played a crucial role in expanding our understanding of regional challenges and opportunities. Additionally, the support and belief from industry frontrunners like the late Mr. K.V. Nageswara Rao (Priya Feeds), Mr. Y. Ravindra Babu (progressive farmer), and Mr. Kothapalli Venkata Surendra (Rani Aqua Feeds) has been instrumental in Marsco's growth.

I want to acknowledge the contributions of my dedicated team. Mr. Prasanta Kumar Satapathy, General Manager of Marsco Nutraceuticals, has been a constant source of drive and innovative thinking. His leadership, alongside the entire Marsco family, ensures that the knowledge and resources we've gained are translated into real, on-the-ground impact for farmers.

From Homemakers to Entrepreneurs

I take immense pride in having introduced women, often the most oppressed among the oppressed, to the world of aquaculture. During my work with rural landless or smallholder farmers, I noticed a gradual drift away from aquaculture due to various practical and desperate reasons. Observing that homemaking women were often in close proximity to the ponds, I saw an opportunity to train them in aquaculture. The prospect of an additional income source that didn't require them to leave their homes was appealing, and they were easily convinced.

My team and I undertook the task of motivating and training these women. Subsequently, NGOs stepped in, providing loans without collateral. We then began monitoring their progress, learning valuable lessons along the way. Our primary intention was to improve their nutrition by increasing the fish production in their homestead ponds. The idea was simple: more fish in the ponds meant more fish on their plates. This initiative not only empowered these women but also contributed to their family's well-being.

In the modest tapestry of my life, every thread tells a story. My entrepreneurial journey with Marsco is a vibrant thread, woven with the colours of passion, resilience, and a deep-seated commitment to empowering the marginalized aqua-farming community. My journey began with a spark - a profound quote by Mahatma Gandhi on the 'Seven Deadly Sins'. This spark ignited a flame that guided me through the labyrinth of commerce and ethics, illuminating my path and shaping my destiny.

The establishment of Marsco was a leap of faith, a right step towards realizing my dream. Despite the challenges and obstacles, I remained steadfast in my commitment to quality and innovation. My tireless efforts to develop low-cost fish farming techniques, my dedication to enhancing farm productivity, and my unwavering resolve to uplift the rural poor are testaments to my enduring spirit.

EoM/

Here are a few suggestions for Blurbs & Excepts

Blurb 1:

"Aquaculture isn't just farming fish, it's farming the future...it wasn't statistics that propelled me... it was witnessing the struggles of individual farmers."

Blurb 2:

"Are business and ethics fundamentally at odds? This is the beginning of my exploration into this question."

Blurb 3:

"Caught between my duty to the bank and the crushing weight of Rebba's predicament...faces of other defaulters are etched in my mind."

Blurb 4:

"Pablo Neruda's words resonated like a clarion call...the realization that true fulfillment stems from following one's dreams."

Blurb 5:

"My journey with Marsco is my way of investing in that future...empowering aqua farmers and building a better future for them and generations to come."

Blurb 6:

"Empowering women...the prospect of an additional income source that didn't require them to leave their homes was appealing."

Excerpt 1:

"I've been privileged to learn from and collaborate with some of the brightest minds in the aquaculture sector. Leading figures like Dr. B.K. Das (ICAR-CIFRI) and Dr. Bimal Prasanna Mohanty (ICAR) have shared their invaluable insights, propelling Marsco's innovations."

Excerpt 2

"CAA approval for those first 20 products, achieving ISO certification…it felt like leaping over the first hurdle. Setting up godowns in Vijayawada, Jajpur, and Balasore – there were times I was that delivery guy."

Excerpt 3

"Being called a 'Startup' still makes me chuckle. A middle-aged guy with grey hairs, out there with the youngsters! But that recognition validated something I believe – it's never too late."

Excerpt 4

"My team knows they can disagree, come up with better solutions than I could. It's about rewarding hard work, showing them a world of possibilities, and giving back with webinars...that's 2000 farmers trained...that's paying back the help I received tenfold."

EoM/

JHELUM

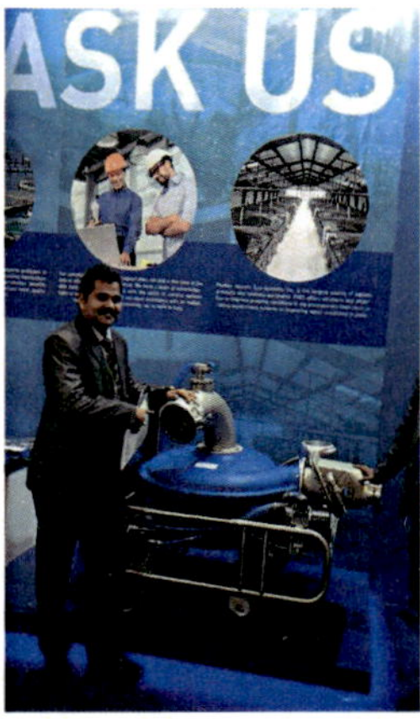
ASK US

Place: Haladipada, Balasore

MARSCO
Marsco Aqua Clinics - Aqua One Centre
National Webinar Series - 10
Disease and Health Management in Freshwater Aquaculture
and
Feed Formulation & Management in Freshwater Aquaculture
Monday, 28th March 2022, 11am to 2pm
Register Now
Free Webinar | e-Certificate will be issued
www.marsco.in
1800 419 9369

MARSCO
Marsco Aqua Clinics - Aqua One Centre
National Webinar Series – 9
Hatchery / Breeding Technologies of Pangasius & Fresh Water Ornamental fish farming
February 18, 2022 | Friday | 11.00 A.M to 2.00 P.M
Dr. S. Ferosekhan
Scientist
ICAR - CIFA
Dr. Ambika P. Nayak
Scientist - KVK
Mr. Mahidhar Bolem
Founder & Director
Marsco Group
REGISTER NOW
Free Webinar | e-Certificate will be issued
www.marsco.in
1800 419 9369

MARSCO
Digital Farming
Wellcome to all
MAXIMIZE YOUR
PROFITS